DATE DUE

Volume 43

Advances
in
Genetics

Volume 43

Advances in Genetics

Edited by

Jeffrey C. Hall
Department of Biology
Brandeis University
Waltham, Massachusetts

Jay C. Dunlap
Department of Biochemistry
Dartmouth Medical School
Hanover, New Hampshire

Theodore Friedmann
Center for Molecular Genetics
    University of California at San
    Diego School of Medicine
La Jolla, California

Francesco Giannelli
Division of Medical and
    Molecular Genetics
United Medical and Dental
    Schools of Guy's and St.
    Thomas' Hospitals
London Bridge, London
United Kingdom

ACADEMIC PRESS
A Harcourt Science and Technology Company

San Diego   San Francisco   New York
Boston   London   Sydney   Tokyo

Academic Press
A Harcourt Science and Technology Company
525 B Street, Suite 1900, San Diego, California 92101-4495, USA
http://www.academicpress.com

Academic Press
Harcourt Place, 32 Jamestown Road, London NW1 7BY, UK
http://www.academicpress.com

International Standard Book Number: 0-12-017643-2

PRINTED IN THE UNITED STATES OF AMERICA
00 01 02 03 04 05 EB 9 8 7 6 5 4 3 2 1

Contents

4 Primary Immunodeficiency Mutation Databases   103

Mauno Vihinen, Francisco X. Arredondo-Vega,
Jean-Laurent Casanova, Amos Etzioni, Silvia Giliani,
Lennart Hammarström, Michael S. Hershfield,
Paul G. Heyworth, Amy P. Hsu, Aleksi Lähdesmäki,
Ilkka Lappalainen, Luigi D. Notarangelo,
Jennifer M. Puck, Walter Reith, Dirk Roos,
Richard Fabian Schumacher, Klaus Schwarz,
Paolo Vezzoni, Anna Villa, Jouni Väliaho,
C. I. Edvard Smith

Contributors

Numbers in parentheses indicate the pages on which the authors' contributions begin.

Ruppa S. Appa (33) Department of Molecular and Medical Pharmacology, UCLA AIDS Institute and Molecular Biology Institute, UCLA School of Medicine, Los Angeles, California 90095

Francisco X. Arredondo-Vega (103) Department of Medicine, Division of Rheumatology, Allergy, and Immunology, and Department of Biochemistry, Duke University Medical Center, Durham, North Carolina 27710

Mark Berneburg (71) MRC Cell Mutation Unit, University of Sussex, Falmer, Brighton, BN1 9RR, United Kingdom

Jean-Laurent Casanova (103) Unite Clinique d'Immunologie et d'Hématologie Pediatriques et Laboratoire INSERM U429, Hopital Necker-Enfants Malades, 149 rue de Sèveres, 75015 Paris, France

Samson A. Chow (33) Department of Molecular and Medical Pharmacology, UCLA AIDS Institute and Molecular Biology Institute, UCLA School of Medicine, Los Angeles, California 90095

Robin A. J. Eady (1) Department of Cell and Molecular Pathology, St. John's Institute of Dermatology, The Guys, Kings College and St. Thomas' Hospital Medical School, London, SE1 7EH, United Kingdom

Amos Etzioni (103) Department Pediatrics, Rambam Medical Center, B. Rappaport School of Medicine, Haifa, Israel

Silvia Giliani (103) Istituto di Medicina Molecolare "Angelo Nocivelli," Department of Pediatrics, University of Brescia, Brescia, Italy

Lennart Hammarström (103) Division of Clinical Immunology, Karolinska Institute at Huddinge Hospital, S-141 86 Huddinge, Sweden

Michael S. Hershfield (103) Department of Medicine, Division of Rheumatology, Allergy, and Immunology, and Department of Biochemistry, Duke University Medical Center, Durham, North Carolina 27710

Paul G. Heyworth (103) Department of Molecular and Experimental Medicine, The Scripps Research Institute, La Jolla, California 92037

Michelle L. Holmes-Son (33) Department of Molecular and Medical Pharmacology, UCLA AIDS Institute and Molecular Biology Institute, UCLA School of Medicine, Los Angeles, California 90095

Amy P. Hsu (103) Genetics and Molecular Biology Branch, National Human Genome Research Institute, National Institutes of Health, Bethesda, Maryland 20892

Alekski Lähdesmäki (103) Division of Clinical Immunology, Karolinska Institute at Huddinge Hospital, S-141 86 Huddinge, Sweden

Ilkka Lappalainen (103) Institute of Medical Technology, FIN-33014 University of Tampere, Tampere, Finland
Department of Biosciences, Division of Biochemistry, PO Box 56, FIN-00014 University of Helsinki, Finland

Alan R. Lehmann (71) MRC Cell Mutation Unit, University of Sussex, Falmer, Brighton, BN1 9RR, United Kingdom

John A. McGrath (1) Department of Cell and Molecular Pathology, St. John's Institute of Dermatology, The Guys, Kings College, and St. Thomas' Hospital Medical School, London, SE1 7EH, United Kingdom

Luigi D. Notarangelo (103) Istituto di Medicina Molecolare "Angelo Nocivelli," Department of Pediatrics, University of Brescia, Brescia, Italy

Jennifer M. Puck (103) Clinical Pathology Department, Warren Magnusen Clinical Center, National Institutes of Health, Bethesda, Maryland 20892

Walter Reith (103) Department of Genetics and Microbiology, University of Geneva Medical School, 1rue Michel-Servet, CH-1211 Geneva 4, Switzerland

Dirk Roos (103) Central Laboratory of the Netherlands Blood Transfusion Service and Laboratory for Experimental and Clinical Immunology, Academic Medical Center, University of Amsterdam, Amsterdam, The Netherlands

Richard Fabian Schumacher (103) Istituto di Medicina Molecolare "Angelo Nocivelli," Department of Pediatrics, University of Brescia, Brecia, Italy

Klaus Schwarz (103) Department of Transfusion Medicine, University of Ulm, Ulm, Germany

C. I. Evard Smith (103) Division of Clinical Immunology, Karolinska Institute at Huddinge Hospital, S-141 86 Huddinge, Sweden
Clinical Research Center, Karolinska Institutet at Huddinge Hospital, S-14186 Huddinge, Sweden

Jouni Väliaho (103) Institute of Medical Technology, FIN-33014 University of Tampere, Tampere, Finland

Paolo Vezzoni (103) Department of Human Genome and Multifactorial Disease, Consiglio Nazionale delle Ricerche, Via Fratelli Cervi 93, 20090 Segrate (Milan) Italy

Mauno Vihinen (103) Institute of Medical Technology, FIN-33014 University of Tampere, Tampere, Finland

Anna Villa (103) Department of Human Genome and Multifactorial Disease, Consiglio Nazionale delle Ricerche, Via Fratelli Cervi 93, 20090 Segrate (Milan), Italy

1

Recent Advances in the Molecular Basis of Inherited Skin Diseases

John A. McGrath* and Robin A. J. Eady
Department of Cell and Molecular Pathology
St John's Institute of Dermatology
The Guy's, Kings College and St Thomas' Hospital Medical School
St. Thomas' Hospital
London, SE1 7EH, United Kingdom

I. Introduction
II. Organization and Composition of the Epidermis and
   Dermal–Epidermal Junction
   A. The Epidermis
   B. Dermal–Epidermal Junction
III. Inherited Skin Disease Mutations
   A. Epidermolysis Bullosa Simplex
   B. Epidermolysis Bullosa with Muscular Dystrophy
   C. Epidermolytic Hyperkeratosis/Bullous Congenital
      Ichthyosiform Erythroderma
   D. Ichthyosis Bullosa of Siemens
   E. Pachyonychia Congenita
   F. Epidermolytic Palmoplantar Keratoderma
   G. Striate Palmoplantar Keratoderma
   H. Skin Fragility/Ectodermal Dysplasia Syndrome
   I. Darier's Disease
   J. Lamellar Ichthyosis
   K. Erythrokeratodermia Variabilis
   L. Progressive Symmetric Erythrokeratodermia

*To whom correspondence should be addressed. Telephone: 44-20-7928-9292 ext.3318. Fax: 44-20-7922-8175. E-mail: john.mcgrath@kcl.ac.uk.

Advances in Genetics, Vol. 43

1

ABSTRACT

Over the last few years the molecular basis of several inherited skin diseases has been delineated. Some discoveries have stemmed from a candidate gene approach using clinical, biochemical, immunohistochemical, and ultrastructural clues, while others have arisen from genetic linkage and positional cloning analyses. Notable advances have included elucidation of specific gene pathology in the major forms of inherited skin fragility, ichthyosis, and keratoderma. These findings have led to a better understanding of the significance of individual structural proteins and regulatory enzymes in keratinocyte adhesion and differentiation. From a clinical perspective, the advances have led to better genetic counseling in many disorders, the development of DNA-based prenatal diagnosis, and a foundation for planning newer forms of treatment, including somatic gene therapy, in selected conditions.   © 2001 Academic Press.

I. INTRODUCTION

In recent years, the skin has proved to be a fertile "hunting ground" for uncovering a wide variety of causative genes underlying several major hereditary skin disorders. In fact, some of these advances have elucidated the molecular basis of diseases which are not skin-restricted and have had much wider implications.

For example, the discovery of mutations in the gap junction protein, connexin 26, as a cause of keratoderma (thickening of the skin of palms and soles) in individuals who also suffer with deafness has opened up a totally new avenue of research into the genetic cause of sensorineural deafness (Kelsell *et al.*, 1997).

The main investigative approaches that have been used to find causative genes in hereditary skin disorders have tended to fall into three main categories. First, the skin lends itself particularly well to the candidate gene approach. This is because the phenotypic or clinical manifestations resulting from gene mutations may be readily examined in a patient's skin. Changes in texture, thickness, color, or fragility in the skin, hair, or nails are generally easy to detect. Further, detailed examination of skin biopsies using electron microscopy and/or immunohistochemical analysis has revealed critical changes in intra- or extracellular ultrastructural components or in the amount and distribution of protein expression. This approach has been particularly valuable in flagging candidate genes in autosomal recessive disorders—for example, in the different forms of epidermolysis bullosa.

Second, reverse genetic approaches, including positional cloning to define the genetic locus and ultimately discover the causative gene, have also had a major role in elucidating genes underlying diseases such as epidermolysis bullosa simplex, basal cell nevus syndrome, and Darier's disease. In the latter two disorders, and as in many autosomal dominant conditions, the candidate gene approach was, on the whole, unhelpful. However, the discovery that mutations in the keratin 14 gene and, later, the keratin 5 gene were the cause of epidermolysis bullosa simplex came via the alternative but complementary route of positional cloning and transgenic mouse technology. As a third line of investigation, transgenic and knockout mouse experiments have also played a major role not only in identifying candidate genes in human genetic skin disorders, but also in providing the means for research on mechanisms in disease and on the development of newer forms of treatment.

Because many of the recent advances in elucidating hereditary skin disorders apply to diseases whose phenotype is entirely or largely manifested in the epidermis or dermal–epidermal junction, this chapter will focus on these two major zones in the skin, instead of attempting to cover the fuller variety of tissues and cell types that are present in the dermal connective tissue and subcutaneous tissue.

II. ORGANIZATION AND COMPOSITION OF THE EPIDERMIS AND DERMAL–EPIDERMAL JUNCTION

A. The epidermis

The epidermis is a stratified squamous epithelium whose principal cells, the keratinocytes, arise in the basal layer as a result of mitosis of the resident stem cells, and migrate through the spinous and granular layers until they reach the most superficial compartment, the cornified cell layer or stratum corneum (Figure 1.1), from which they are eventually shed as dead, anucleate squames. This highly ordered process, starting with the cycling of stem cells in the basal, germinative compartment and ending with dead or dying cells higher up, is known as terminal differentiation.

In order for the epidermis effectively to fulfill one of its major functions in establishing, maintaining, and repairing a barrier between the living cells of

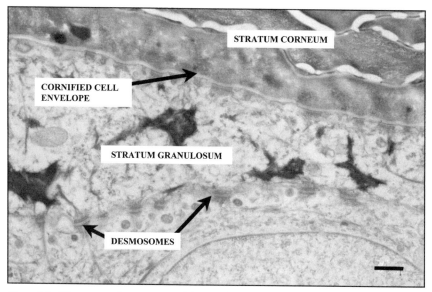

Figure 1.1. Ultrastructural appearances of the superficial part of normal epidermis depicting desmosomes, stratum granulosum (granular cell layer), cornified cell envelope and stratum corneum. (Bar = 2 μm.)

the skin and the outside environment, it is necessary to have a strictly controlled and concerted action by the population of keratinocytes, which, in turn, has to be securely anchored to the basement membrane and through this to the underlying dermis.

1. Intermediate filament proteins

Intermediate filament (IF) proteins are members of a superfamily of evolutionary conserved genes and proteins (Lazarides, 1980; Irvine and McLean, 1999). Several distinct types of IF have been described on the basis of their gene structure, amino acid sequence, and tissue expression. The different subtypes comprise vimentin, glial filament acidic protein, neurofilaments, desmin, peripherin, nuclear lamins, and keratins. The keratins are the characteristic 10-nm-diameter intermediate filament of all epithelial cells.

2. Keratins

Keratins form two main groups, type I and type II, based on their molecular size, charge, and gene structure. The acidic (pH = 4–6), type I keratins include 11 epithelial-type keratins K9-K20 (K18 being an exception) and the Ha trichocyte keratins. Type II keratins are neutral/basic (pH = 6–8) and comprise eight members, K1–K8, and the Hb trichocyte keratins. Type I keratin genes are present in a cluster on chromosome 17q21 and type II keratin genes form a second cluster on chromosome 12q13.

Keratin IF formation involves the dimerization of a keratin pair with one of the pair having a type I keratin, and the other, type II. In normal basal epidermis, the predominant keratins are K5 and K14. In the suprabasal spinous layer, K5 and K14 are downregulated and K1 and K10 are preferentially expressed. Some keratins are site-restricted; for example, K2e is located within superficial interfollicular epidermis, K9 is limited to palmoplantar skin, and K19 is found in basal keratinocytes of the hair follicle bulge region.

3. Cornified cell envelope

The cornified cell envelope is an insoluble, tough protective skin barrier formed beneath the cell membrane during terminal differentiation. Its rigidity is derived from the cross-linking of a number of structural proteins including involucrin, loricrin, SPRRs, elafin, cystatin A, S100 family proteins, and some components of desmosomes, in a complex process catalyzed by sulfydryl oxidase and transglutaminases (Ishida-Yamamoto and Iizuka, 1998). Considerable insight into the roles played by components of the cornified envelope has been derived from dominant-negative and knockout mice studies as well as autosomal dominant and recessive inherited human skin disorders. Such studies highlight the presence of a degree of functional redundancy among the cornified envelope components. For example, −/− loricrin mice fail to show any sustained phenotypic abnormalities (De Viragh et al., 1997), nor do mice that overexpress loricrin (Yoneda and Steinert, 1993), whereas mice with dominant-negative mutations display features similar to the human disorder, Vohwinkel's syndrome (Suga et al., 1999).

4. Intraepidermal junctions

Adjacent keratinocytes are linked by a series of different junctional complexes which help regulate cell adhesion and cell–cell communication. The principal junctional mechanisms comprise desmosomes, adherens junctions, gap junctions, and tight junctions (Figure 1.2). Desmosomes are circular membrane domains, 0.1–0.5 μm in diameter, and are present in nearly all epithelia. They consist of calcium-dependent transmembranous glycoproteins, the cadherins—desmocollins and desmogleins—as well as other components such as desmoplakin and plakoglobin that link the desmosomal plaque to the keratin filament cytoskeleton (Kowalczyk et al., 1999; North et al., 1999). Expression of certain desmosomal components is site-restricted. For example, plakophilin 1 is located predominantly in desmosomes in terminally differentiating keratinocytes. Detailed transgenic

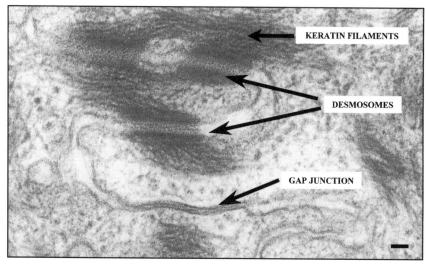

Figure 1.2. Ultrastructural features of desmosomes and gap junctions. Desmosomes appear as compact, multilaminated, electron-dense adhesion junctions associated with insertion of keratin intermediate filaments. Gap junctions are less conspicuous but may be seen in proximity to desmosomes. (Bar = 0.1 μm.)

mouse studies and naturally occurring human desmosome gene mutations have helped reveal critical roles and protein-binding domains of individual desmosone components in maintaining epithelial integrity. The main component of gap junctions are the connexins, the precise relevance of which is starting to emerge from mutational analysis in patients with different forms of nonsyndromic sensorineural deafness and/or keratoderma (Kelsell *et al.*, 1997).

B. Dermal–epidermal junction

At the interface between the epidermis and dermis is a compact, 200-nm-wide zone that contains numerous adhesive proteins and glycoproteins (Figures 1.3 and 1.4). Two main adhesion complexes are recognized, hemidesmosomes and focal contacts, which provide attachment points for the keratin and actin filament networks, respectively.

1. Hemidesmosomes

Ultrastructurally, hemidesmosomes are electron-dense studlike structures located along the basal pole of basal keratinocytes. They comprise an inner plaque that

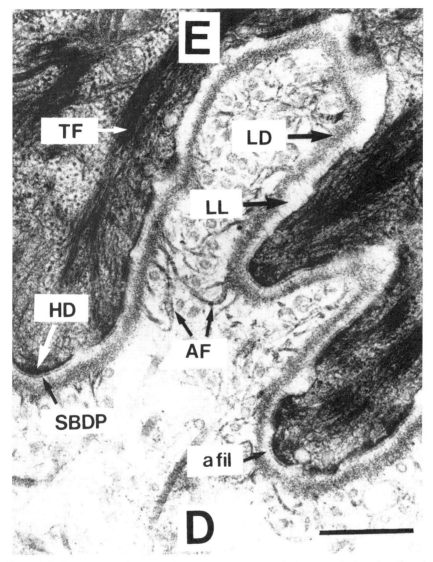

Figure 1.3. Transmission electron microscopy appearances of the normal dermal–epidermal junction. E = epidermis; D = dermis; TF = tonofilaments; HD = hemidesmosome; SBDP = subbasal dense plate; LL = lamina lucida; LD = lamina densa; a fil = anchoring filaments; AF = anchoring fibrils. (Bar = 0.25 μm.)

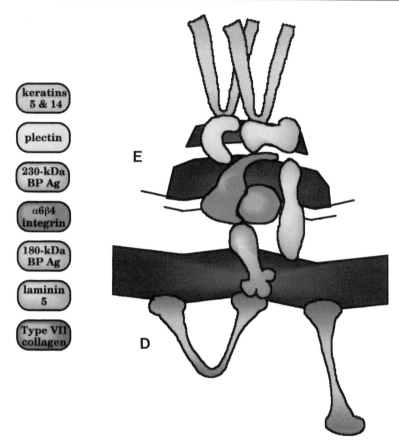

keratins 5 & 14

plectin

230-kDa BP Ag

α6β4 integrin

180-kDa BP Ag

laminin 5

Type VII collagen

E

D

Figure 1.4. Schematic representation of the adhesive network of proteins and glycoproteins that constitute the dermal–epidermal junction. These macromolecules form an interactive bridge between the keratin filaments of basal keratinocytes and collagen fibres in the dermis. E = epidermis; D = dermis.

is closely associated with the keratin filaments and an outer plaque abutting the plasma membrane. Key components of hemidesmosomes include the 230-kDa bullous pemphigoid antigen (BPAG1), plectin, type XVII collagen (BPAG2, also known as the 180-kDa bullous pemphigoid antigen), and α6β4 integrin. Plectin and BPAG1 belong to the plakin family of proteins and bind directly to keratin filaments through domains that have homology to the desmosomal protein, desmoplakin. Adhesion to the plasma membrane is established through the binding of plectin and BPAG1 to the transmembranous proteins, type XVII collagen, and α6β4 integrin (Borradori and Sonnenberg, 1999).

2. Lamina lucida/lamina densa

In the immediate extracellular space the dermal–epidermal junction comprises an electron-lucent layer (lamina lucida) and an electron-dense layer (lamina densa), which contain a number of proteins and glycoproteins including laminins 5 and 6, nidogen-entactin, heparan sulfate proteoglycan, and type IV collagen. The lamina lucida is spanned by fine (3- to 4-nm-diameter) threadlike structures known as anchoring filaments. These filaments contain the extracellular domains of type XVII collagen and $\alpha 6\beta 4$ integrin as well as laminin 5 and 6 at the lamina lucida/lamina densa interface. Type IV collagen represents the major structural component of the lamina densa.

3. Anchoring fibrils

Just subjacent to the lamina densa is a zone of anchoring fibrils which are composed of type VII collagen. Most anchoring fibrils form loops within the superficial dermis that are anchored by the amino-terminus of type VII collagen within the lamina densa. The looping fibrils are traversed by interstitial collagen fibrils (mainly type I and III collagen), thereby securing adhesion to the dermis. Type VII collagen also binds to laminin 5, thus adding to the maromolecular adhesive network. Indeed, the structural integrity of the dermal–epidermal junction depends on the complex arrangement of protein–protein interactions that extend from within basal keratinocytes into the dermis. Disruption of this skin region through inherited mutations in the corresponding genes leads to skin fragility and the group of disorders known as epidermolysis bullosa (Fine *et al.*, 2000; Marinkovich, 1999; Mellerio, 1999).

III. INHERITED SKIN DISEASE MUTATIONS

The molecular basis of several structural and regulatory proteins relevant to keratinocyte cell biology has recently been elucidated. A number of significant discoveries are highlighted below.

A. Epidermolysis bullosa simplex

Epidermolysis bullosa simplex (EBS) is a heterogeneous group of inherited skin fragility disorders; most are autosomal dominant, although autosomal recessive transmission is recognized rarely. EBS involves fragility of basal keratinocytes and was the first identified human keratin disease (Figure 1.5a). The most common, and mildest, form of EBS is the Weber-Cockayne subtype (MIM131800), which usually comprises mild blistering of the hands and feet, worse during summer

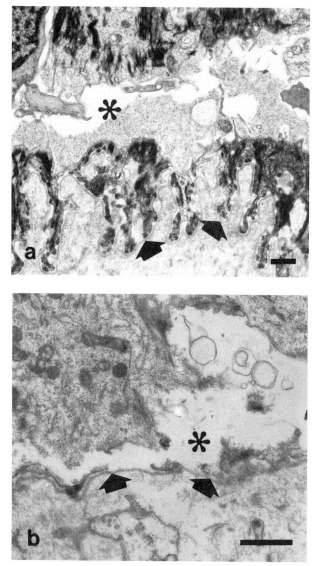

Figure 1.5. Different ultrastructural planes of cleavage at or close to the dermal–epidermal junction in various forms of epidermolysis bullosa (EB) (asterisk = level of blister formation; arrows = lamina densa). (a) In EB simplex (Dowling-Meara), cleavage occurs through the basal keratinocytes. (b) In EB simplex associated with muscular dystrophy, tissue separation occurs just above the plasma membrane of basal keratinocytes. (c) In junctional EB, the plane of cleavage is within the lamina lucida. (d) In dystrophic EB, the level of blister formation is just below the lamina densa. (Bar = 10 μm.)

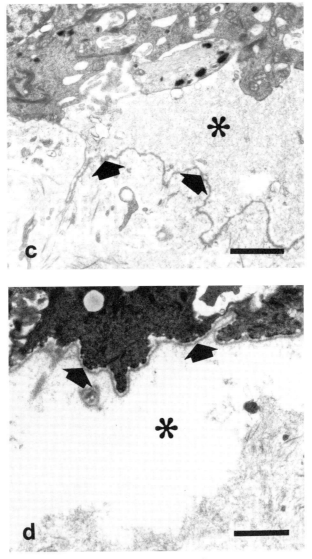

Figure 1.5. (*Continued*)

months. The second most common form of EBS is the Dowling-Meara variant (MIM 131760). This condition may be associated with severe neonatal erythema and skin blistering, but gradually the blisters tend to occur in clusters and, with time, most patients develop hyperkeratosis of the palms and soles. Most forms of EBS result from mutations in keratin 5 or 14. Initial clues to the inherent gene pathology arose from immunoelectron microscopy (Ishida-Yamamoto *et al.*, 1991) and transgenic mouse studies (Vassar *et al.*, 1991), but a large number of pathogenic mutations in K5 or K14 has been identified subsequently (Bonifas *et al.*, 1991; Coulombe *et al.*, 1991; Lane *et al.*, 1992). In severe forms of EBS, the mutations tend to occur within the highly conserved helix boundary motifs, with a mutation hotspot in a critical arginine residue in the IA domain (R125) delineated in many affected individuals. In milder forms of EBS, the mutations are sited away from the helix initiation and termination peptides (Rugg *et al.*, 1993). A rare subtype of autosomal dominant EBS associated with mottled pigmentation (MIM131960) has been shown to be due to a heterozygous missense mutation in keratin 5, P24L, in the nonhelical VI domain of K5 (Uttam *et al.*, 1996). Most autosomal recessive forms of EBS are associated with functional knockout mutations resulting in complete ablation of keratin 14 (Rugg *et al.*, 1994). Interestingly, the phenotype in affected humans is milder than that in mice with K14 ablation through gene targeting, which die within a few months because of esophageal fragility. It is possible that in humans there is some functional compensation through increased expression of K15 (Lloyd *et al.*, 1995).

B. Epidermolysis bullosa with muscular dystrophy

Epidermolysis bullosa with late-onset muscular dystrophy (EB-MD) (MIM 226670) is an autosomal recessive disorder. Skin blistering may be mild, although mucosal involvement may cause significant morbidity (McLean *et al.*, 1996; Smith *et al.*, 1996; Mellerio *et al.*, 1997). The major concern, however, is the development of muscular dystrophy, usually in the second or third decade. The molecular pathology involves mutations in plectin, a component of hemidesmosomes found in several epithelial and mesenchymal tissues, including striated muscle (Wiche, 1998). Skin fragility occurs within basal keratinocytes just above the plasma membrane through the inner plaque of hemidesmsomes (Figure 1.5b). The most severe cases are associated with complete ablation of plectin with nonsense mutations on both alleles, while milder cases may result from in-frame deletions or insertions (Rouan *et al.*, 2000).

C. Epidermolytic hyperkeratosis/bullous congenital ichthyosiform erythroderma

Epidermolytic hyperkeratosis (EH), also known as bullous congenital ichthyosiform erythroderma (BCIE) (MIM 113800), is an autosomal dominant disorder characterized by neonatal erythema and blistering which then evolves into generalized hyperkeratosis (Figure 1.6). Clues to the pathogenesis of the disorder also came from immunoultrastructural (Ishida-Yamamoto *et al.*, 1992) and transgenic mouse studies (Fuchs *et al.*, 1992) which implicated keratins 1 and 10 in its

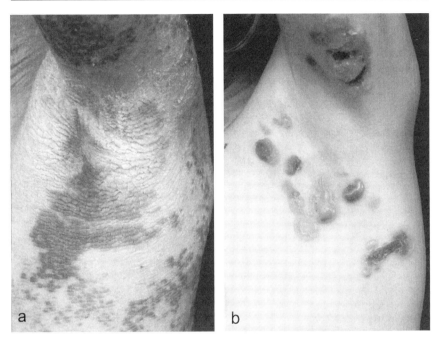

Figure 1.6. Clinical consequences of mutations in keratin genes. (a) Mutations in the genes encoding keratin 1 or 10 result in epidermolytic hyperkeratosis/bullous congenital ichthyosiform erythroderma. This picture shows axillary hyperkeratosis and skin scaling. (b) Mutations in the genes for keratin 5 or 14 lead to epidermolysis bullosa simplex. This picture illustrates clustered blistering in the axilla of a patient with the Dowling-Meara variant.

pathogenesis, and subsequently, a number of mutations in the genes for K1 and K10 have been reported (Rothnagel *et al.*, 1992). Most mutations are missense and are clustered at the ends of the central helical rod domains. Mutations in the 2B domain of K1 may be associated with an atypical presentation of cyclic ichthyosis with epidermolytic hyperkeratosis (Sybert *et al.*, 1999). Mutations in K1 may be associated with more severe palmoplantar hyperkeratosis compared to K10 mutations, but this is not absolute. Epidermolytic hyperkeratosis is also the only human keratin disease to exhibit mosaicism (either keratin 1 or 10), which may be confined to the skin (linear EH) or may also involve gonadal mosaicism (Paller *et al.*, 1994). Patients with nevoid forms of EH, therefore, may be at risk of having offspring with generalized disease (DiGiovanna and Bale, 1994).

D. Ichthyosis bullosa of Siemens

Ichthyosis bullosa of Siemens (IBS) (MIM 146800) represents a form of epidermolytic hyperkeratosis that is clinically milder than that associated with K1/K10 mutations. Most of the hyperkeratosis is confined to the flexures and usually there is no erythroderma. In addition, IBS shows a characteristic "Mauserung" molting of the outer layers of the epidermis. Occasionally, however, there may be clinical overlap between EH and IBS. Histologically, the tonofilament disruption is more superficial than in EH. The disorder results from mutations in the type II basic keratin, K2e. Mutational analyses have suggested hotspot mutations underlying a disproportionate number of cases (Basarab *et al.*, 1999; Suga *et al.*, 2000).

E. Pachyonychia congenita

Pachyonychia congenita (PC) encompasses a group of autosomal dominant disorders with prominent hypertrophic nail dystrophy. Two main subtypes are recognized: PC-1 (MIM 167200) and PC-2 (MIM 167210). PC-1 is associated with severe focal keratoderma, whereas PC-2 is accompanied by multiple steatocytes which appear at pregnancy, as well as mild focal keratoderma, pili torti, and natal teeth. Both subtypes may be associated with variably penetrable features such as angular chelosis, follicular keratosis, hoarseness, and oral leucokeratosis. In PC-1, pathogenic mutations have been delineated in keratin 16 and its expression partner, K6a (Bowden *et al.*, 1995; McLean *et al.*, 1995; Smith *et al.*, 1998a). For PC-2 the corresponding mutations occur in keratin 17 and K6b (McLean *et al.*, 1995; Covello *et al.*, 1998, Smith *et al.*, 1998b). Nevertheless, considerable intra- and interfamilial variation in phenotype is well recognized. For example, K16 mutations may result in focal keratoderma in isolation and K17 mutations may be found in patients with steatocystocytoma multiplex (Covello *et al.*, 1998).

F. Epidermolytic palmoplantar keratoderma

Epidermolytic palmoplantar keratoderma (EPPK) (MIM 144200) is an autosomal dominant disorder associated with hyperkeratosis confined to the palms and soles (Figure 1.7). The skin thickening often has a waxy appearance and is surrounded by an erythematous margin. EPPK is caused by mutations in the type I acidic keratin K9 (Reis *et al.*, 1994). Most mutations are missense and all have been sited in the 1A domain. Unlike most other keratins, the expression partner for K9 is not known, although a K1 isoform remains a likely candidate. Nevertheless, no type II keratin gene mutations have yet been described in EPPK.

G. Striate palmoplantar keratoderma

Striate palmoplantar keratoderma (SPPK) (MIM 148700) is a distinct form of autosomal dominant hyperkeratosis. As the name implies, the disorder usually comprises linear hyperkeratotic changes on the palms and palmar aspects of the

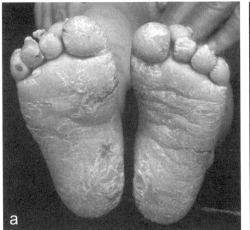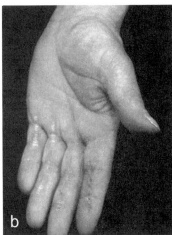

Figure 1.7. Clinical consequences of mutations in genes encoding structural components of desmosomes. (a) Hyperkeratosis and fissuring of the soles and nail dystrophy in a patient with
autosomal recessive null mutations in plakophilin 1. (b) Linear hyperkeratosis on the palmar aspect of the fingers in an individual with autosomal dominant striate palmoplantar
keratoderma resulting from haploinsufficiency of desmoplakin.

fingers. Similar changes may be present on the soles, but the changes are often more
focal. Skin biopsy examination may show abnormalities in keratinocyte adhesion,
and haploinsufficiency mutations in two different components of desmosomes,
desmoplakin and desmoglein-1, have been described (Armstrong et al., 1999;
Rickman et al., 1999; Whittock et al., 1999a). Studies in a number of families
suggest that there may be further genetic heterogeneity and that other factors,
including environmental trauma and modifying genes, may all contribute to the
phenotype of an affected individual (Kelsell and Stephens, 1999; Whittock et al.,
1999a).

H. Skin fragility/ectodermal dysplasia syndrome

A further inherited abnormality of desmosomes recently has been reported in
a previously unrecognized autosomal recessive disorder comprising features of
skin fragility and ectodermal dysplasia (MIM 604536). Affected individuals had
trauma-induced skin blistering and/or crusting as well as nail dystrophy, loss of
hair, and a painful palmoplantar keratoderma with skin cracking (McGrath et al.,
1997) (Figure 1.7). The molecular pathology involves total ablation of the protein
plakophilin 1, which has a vital role in securing keratinocyte adhesion during
terminal differentiation (McGrath et al., 1999a). Patients' skin biopsies showed

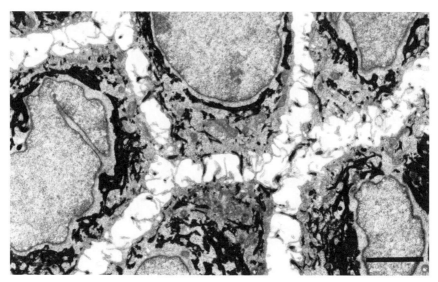

Figure 1.8. Transmission electron micrograph showing loss of keratinocyte adhesion in skin from a patient with complete ablation of plakophilin 1. There is a widening of spaces between adjacent keratinocytes, a reduced number of hypoplastic desmosomes, and perinuclear compaction of the keratin filament network. (Bar = 2 μm.)

loss of adhesion between keratinocytes through the spinous layer (Figure 1.8) as well as a redistribution of labeling for desmoplakin, providing evidence for a close interaction among desmoplakin, plakophilin 1, and the keratin intermediate filament network in maintaining adhesion between adjacent keratinocytes.

I. Darier's disease

Darier's disease (MIM 124200) is an autosomal dominant disorder with abnormalities in keratinocyte adhesion and keratinization (Figure 1.9). Recent years have seen a number of genetic linkage and positional cloning studies which have gradually narrowed the disease gene interval. In 1999, the disorder was finally shown to involve pathogenic mutations in an intracellular calcium pump gene, *ATP2A2*, which encodes the sacro/endoplasmic reticulum Ca^{2+}-ATPase type 2 isoform (SERCA2) which is highly expressed in keratinocytes (Sakuntabhai *et al.*, 1999). These studies showed that the histological changes seen in desmosomes in Darier's diesease are secondary to abnormalities in calcium signaling within keratinocytes. A similar autosomal dominant disorder, Hailey-Hailey disease (MIM 169600) (Figure 1.9), subsequently has also been shown to harbor mutations in a further intracellular calcium pump gene, *ATP2C1* (Hu *et al.*, 2000). Collectively,

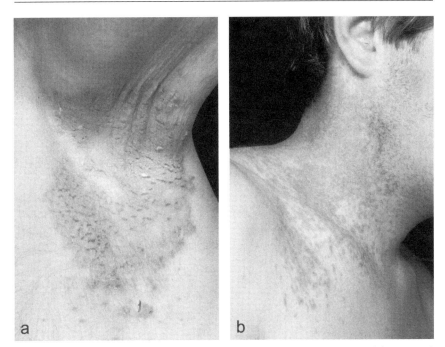

Figure 1.9. Clinical consequences of mutations in intracellular calcium pump genes. (a) Autosomal dominant mutations in the *ATP2C1* gene underlie Hailey-Hailey disease as shown in this patient with erythema and erosions in the axilla. (b) Autosomal dominant mutations in the *ATP2A2* gene give rise to Darier's disease. This picture shows confluent warty erythematous papules and superficial erosions on the neck of an affected individual.

these findings provide new insights into some of the regulatory processes that govern keratinocyte adhesion and differentiation.

J. Lamellar ichthyosis

Lamellar ichthyosis (LI) (MIM 242100) is an autosomal recessive disorder characterized by generalized scaliness and variable redness of the skin. At birth, however, the disorder may show considerable clinical overlap with other congenital ichthyoses, including Netherton syndrome, Sjögren-Larsson syndrome, Rud syndrome, and some of the trichothiodystrophies (Hennies *et al.*, 1998). In some cases of lamellar ichthyosis, genetic linkage has been demonstrated to 14q11 and pathogenic mutations have been shown in the transglutaminase 1 gene which disrupt the function of keratinocyte transglutaminase enzymes and thereby perturb the normal process of cornification and desquamation (Huber *et al.*, 1995;

Russell *et al.*, 1995). However, not all cases of LI map to 14q11: evidence for a second locus at 2q33-35 and possible further loci have been reported (Parmentier *et al.*, 1996, 1999). Thus, LI is a genetically heterogenous disorder.

K. Erythrokeratodermia variabilis

Erythrokeratodermia variabilis (EKV) (MIM 133200) is usually an autosomal dominant disorder, although rare autosomal recessive pedigrees have been described (Armstrong *et al.*, 1997). The disorder is characterized by stable and transient edematous and hyperkeratotic plaques. There are migratory patches of erythema which spread over the trunk and limbs and which may last minutes, days, or months. Pruritus may be a feature (Papadavid *et al.*, 1998). In addition, affected individuals may have nonmigratory fixed yellow-brown scaly plaques. The disorder has been mapped to 1p34-p35 and mutations have been identified in the *GJB3* gene which encodes the gap junction protein, connexin 31 (Richard *et al.*, 1998). This connexin, therefore, appears to have a vital role in intercellular communication with regard to epidermal differentiation.

L. Progressive symmetric erythrokeratodermia

Progressive symmetric erythrokeratodermia (PSEK) is an autosomal dominant disorder characterized by symmetrical red scaly plaques on the knees, buttocks, and groin. The face may be involved, and most patients have palmoplantar hyperkeratosis and pseudoainhum (constriction bands on fingers/toes). The disorder does not have the transient migratory erythema seen in EKV, although clinical overlap of the two disorders in the same family has been described. Mutations in the cornified envelope loricrin in one PSEK family have been described (Ishida-Yamamoto *et al.*, 1997). The mutations resulted in a delayed termination codon in loricrin and abnormal nuclear retention of loricrin with compromised cornified envelope formation. Similar types of mutation have been reported in the ichthyotic variant of Vohwinkel's syndrome.

M. Vohwinkel's syndrome

Vohwinkel's syndrome (VS) (MIM 124500) is an autosomal dominant disorder associated with a papular and honeycomb keratoderma, starfish-like acral keratoses, and constriction of the digits leading to autoamputation. The presence or absence of superimposed ichthyosis or deafness is important, since the disorder is genetically heterogeneous (Korge *et al.*, 1997). The first mutation described in VS was in the variant associated with ichthyosis. Affected individuals had a frameshift mutation in the loricrin gene (Maestrini *et al.*, 1996). In other families with VS, with deafness as an integral feature—as described in Vohwinkel's original description—pathogenic mutations have been delineated in the *GJB2* gene, encoding the gap junction protein, connexin 26 (Maestrini *et al.*, 1999). Mutations in connexin 26 are known to also underlie other examples of autosomal recessive or dominant deafness (Kelsell *et al.*, 1997). As far as the molecular pathology of VS is concerned, "classical" VS results from connexin 26 mutations, whereas the ichthyotic variant is due to mutations in loricrin (Ishida-Yamamoto *et al.*, 1998; Korge *et al.*, 1997).

N. Papillon-Lefèvre syndrome

Papillon-Lefèvre syndrome (MIM 245000) is an autosomal recessive disorder affecting skin and oral epithelium. The most prominent feature is severe periodontitis resulting in premature tooth loss. Other skin signs include scaling over the elbows and knees and palmoplantar hyperkeratosis. The locus for the disorder has been mapped to 11q14-q21 (Fischer *et al.*, 1997), and mutations have been identified in the gene encoding the lysosomal protease cathepsin C (Toomes *et al.*, 1999). The mutations comprised missense, splice site, and nonsense mutations resulting in almost total loss of cathepsin C activity in affected individuals. Cathepsin C has an essential role in the activation of granule serine proteases expressed in myeloid and lymphoid bone marrow cells. Loss of activity may compromise a number of immune and inflammatory processes, including phagocytic destruction of bacteria. However, the relevance of this lysosomal protease to epithelial differentiation and desquamation is not clear. Papillon-Lefèvre syndrome is the second genodermatosis involving lysosomal dysfunction to be reported: Chediak-Higashi syndrome (MIM 214500), whose clinical features consist of periodontitis as well as immune deficits, and abnormalities of pigmentation, blood clotting, and neurological function, has pathogenic mutations in the human equivalent of the mouse beige gene.

O. Monilethrix

Monilethrix (MIM 158000) is an autosomal dominant disorder affecting hair shafts. There is alopecia, beading of hair, and easy fracturing and weathering. Other features may include perifollicular hyperkeratosis and nail abnormalities including koilonychia and brittle nails. Linkage has been established to the trichocyte keratins within the type II keratin gene cluster on 12q13, and mutations have been reported in the trichotype keratins hHb6 and hHb1 (Winter *et al.*, 1997a, 1997b). Monilethrix exhibits a considerable degree of phenotypic variation within families, consistent with additional genetic or environmental factors influencing disease expression (Korge *et al.*, 1999).

P. Congenital atrichia

Congenital atrichia with papular lesions (MIM 209500) is an autosomal recessive disorder characterized by universal congenital alopecia and keratinous skin cysts. Affected individuals may have relatively normal hair at birth, which is then shed during the first few months of life. Genetic linkage studies mapped the disorder to 8p21-22, and mutations were subsequently identified in the human homolog of the mouse hairless gene (Ahmad et al., 1998a). Missense, splice site, frameshift, and nonsense hairless gene mutations have all been documented, although precise genotype–phenotype correlation has yet to be determined (Ahmad et al., 1998b, 1999; Cichon et al., 1998; Kruse et al., 1999). The hairless protein is a transcription factor with a putative zinc finger domain which is disrupted by some of the mutations reported. Clearly, it has a key role in hair development. Another form of inherited alopecia, the autosomal dominant disorder Marie Unna hypotrichosis, maps to a similar locus, although hairless mutations have been excluded in several affected families (van Steensel et al., 1999).

Q. Junctional epidermolysis bullosa

Clinically, junctional epidermolysis bullosa (JEB) is a heterogenous group of autosomal recessive disorders. Severe forms are lethal in the neonatal period or early infancy. Such cases (Herlitz JEB) (MIM 226700) are characterized by ultrastructural evidence of skin fragility within the lamina lucida and morphological alterations in hemidesmosomes/anchoring filaments (Figure 1.5c). At a molecular level, there is total ablation of one of the polypeptide chains of laminin 5 (α3, β3, or γ2), encoded by the genes LAMA3, LAMB3, and LAMC2, respectively (Aberdam et al., 1994; Pulkkinen et al., 1994a, 1994b; Kivirikko et al., 1996). Less severe cases (or non-Herlitz JEB) may result from less disruptive laminin 5 mutations, such as splice site or missense alterations (McGrath et al., 1996), although assessment of the consequences of the mutations on RNA processing may be vital for the accuracy of genetic counseling in some cases (McGrath et al., 1999b) (Figure 1.10). Likewise, other cases of non-Herlitz JEB may result from null mutations in type XVII collagen, also known as the 180-kDa bullous pemphigoid antigen (McGrath et al., 1995). These mutational studies highlight the relative contributions of laminin 5 and type XVII collagen to epidermal–dermal adhesion (Marinkovich 1999). Mutations in the genes encoding the α6β4 integrin (ITGA6 or ITGB4) underlie a distinct variant of JEB associated with pyloric atresia (MIM 226730) (Vidal et al., 1995; Pulkkinen et al., 1997; Ruzzi et al., 1997). Complete ablation of one of these integrins leads to a severe, usually lethal, phenotype, whereas missense or splice site mutations may result in a milder disease, often with relatively trivial skin blistering (Mellerio et al., 1998a). Thus, the clinical diversity in JEB is paralleled by considerable molecular heterogeneity, with

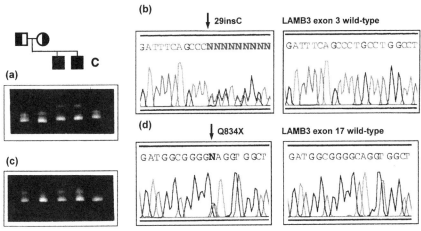

Figure 1.10. Heterozygosity for frameshift/nonsense mutations in the laminin 5 gene, LAMB3, in patients with junctional epidermolysis bullosa (JEB). (a) Heteroduplex analysis of PCR products spanning exon 3 reveals bandshifts in all samples which are similar in the paternal and control, C, DNA, but different in the maternal DNA, which also differs from the bandshifts detected in the JEB individuals. (b) Sequencing of DNA from the mother and the two affected individuals demonstrates a heterozygous frameshift mutation, 29insC. This mutation was not present in the paternal or control DNA samples. However, a silent threonine polymorphism 109 bp downstream from the site of the mutation (138 C-to-T, AC**C** to AC**T**) was detected in DNA from the father, both patients, and the control DNA, but not in the mother's DNA. These findings account for the bandshifts seen in the paternal and control DNA and also for the differences in the type of bandshift seen in the mother and the JEB individuals. (c) Heteroduplex analysis of PCR products spanning exon 17 of LAMB3 reveals bandshifts in DNA from the father and two affected individuals but only homoduplex bands in the maternal and control samples. (d) Sequencing of PCR products displaying heteroduplexes shows a heterozygous C-to-T transition at nucleotide 2504 that changes a glutamine residue (**C**AG) to a nonsense codon (**T**AG). The mutation is designated Q834X. Sequencing of the maternal and control DNA revealed wild-type sequence only. Thus the affected individuals are compound heterozygotes for the mutations 29insC/Q834X in LAMB3. Assessment of RNA revealed that evidence for partial rescue of mutant transcripts containing the mutation Q834X through in-frame exon skipping, thereby moderating the severity of the phenotype (see McGrath et al., 1999b).

pathogenic mutations in six different genes described thus far (Uitto et al., 1997; Mellerio, 1999).

R. Dystrophic epidermolysis bullosa

Dystrophic epidermolysis bullosa (DEB) may be inherited as an autosomal dominant or autosomal recessive blistering skin disease (Figure 1.11). The disorder is associated with a plane of skin fragility below the lamina densa and

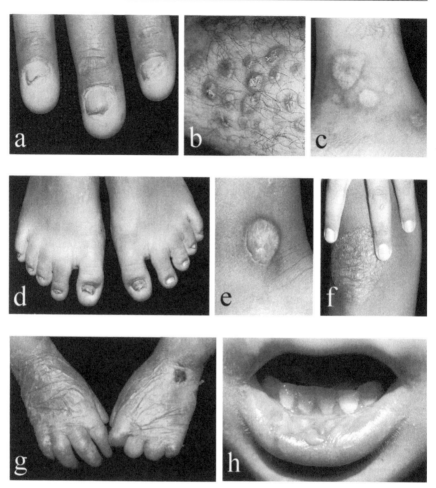

Figure 1.11. Clinical appearances of autosomal dominant and recessive dystrophic epidermolysis bullosa (DEB). In this family, the father has (a) nail dystrophy, (b) blisters and prurigo-like lesions, and (c) scarring. Similar changes are seen in his daughter (d–f), but in his son there is (g) more extensive scarring of the hands and (h) oral erosions and scarring. The father and daughter have dominant DEB resulting from a deletion mutation in the type VII collagen gene (COL7A1), 6863del16, that leads to exon skipping within the triple helix of the anchoring fibril protein, type VII collagen. However, the son has recessive DEB having inherited this mutation *in trans* with a splice-site mutation in COL7A1, 425 A>G, from his clinically normal mother. The carrier frequency for recessive DEB mutations is approximately 1 in 250 individuals, thus overlap of dominant and recessive diseases within the same family is a rare occurrence. Nevertheless, in other families, mutational analysis of COL7A1 may be extremely helpful in determining mode of inheritance in cases of DEB of moderate clinical severity.

ultrastructural changes in anchoring fibril morphology (Bruckner-Tuderman, 1999) (Figure 1.5d).

Immunohistochemical labeling of skin sections shows attenuated or absent staining for type VII collagen, particularly in the more severe clinical subtypes (McGrath et al., 1993). All forms of DEB result from mutations in the type VII collagen gene, COL7A1 (Christiano et al., 1993; Hilal et al., 1993). Linkage studies have excluded the collagenase gene locus as being of primary pathogenic significance (Hovnanian et al., 1991). Dominant forms of DEB (MIM 131700) are usually mild, with blistering and scarring usually restricted to bony prominences. The characteristic dominant DEB mutation in COL7A1 is a glycine substitution within the collagenous triple helix. The most severe forms of recessive DEB (Hallopeau-Siemens; MIM 226600) show extensive blistering and scarring affecting skin and mucous membranes. In these cases, the molecular abnormalities typically involve nonsense mutations on both COL7A1 alleles. The spectrum of clinical severity in other forms of recessive DEB is reflected in a range of COL7A1 pathology, including missense, splice site, and delayed termination codon mutations (Uitto et al., 1997; Whittock et al., 1999b).

IV. TRANSLATIONAL BENEFITS OF MUTATION DISCOVERIES

Discovering the precise molecular abnormality in a patient with an inherited skin disease has several immediate benefits, with particular relevance to genetic counseling, prenatal diagnosis, and the prospect of newer forms of treatment, including somatic gene therapy. In this section, these advances are illustrated with respect to epidermolysis bullosa.

A. Genetic counseling

As in several other inherited disorders, there are a number of genodermatoses in which it is sometimes difficult to distinguish clinically between autosomal dominant and autosomal recessive inheritance. For example, in the case of a child with a mild form of dystrophic epidermolysis bullosa (DEB) and clinically normal parents, the distinction between recessive and de novo dominant disease is fundamental to the basis of genetic counseling (Hashimoto et al., 1999). Extensive mutational analyses have demonstrated that most such cases are in fact autosomal recessive in nature, although a number of exceptions have been noted (Whittock et al., 1999b). Counseling can be further complicated by the fact that glycine substitutions in type VII collagen, the paradigm of dominant DEB mutations, may also be inherited in a recessive fashion, having no clinical features when present with a second normal COL7A1 allele (although subtle toe nail dystrophy should always be searched for) and only manifesting as an inherited blistering skin disease when inherited in trans with a second mutated allele (Dunnill et al., 1996;

Hammami-Hauasli *et al.*, 1998; Terracina *et al.*, 1998; Shimizu *et al.*, 1999). Mutational analysis may have an even more fundamental role in determining prognosis. For example, in a neonate diagnosed with junctional epidermolysis bullosa (JEB), demonstration of null mutations in a laminin 5 gene typically underlies the severe Herlitz form of the disorder, while null mutations in type XVII collagen indicate a relatively milder form of non-Herlitz JEB. Furthermore, identification of certain specific mutations may have particular prognostic value. For example, delineation of the mutation E210K in the *LAMB3* (laminin 5) gene, in combination with a nonsense mutation on the other *LAMB3* allele, predicts development of non-Herlitz disease with atrophic changes on the shins, dental enamel hypoplasia, nail dystrophy, and patchy alopecia (Mellerio *et al.*, 1998b). Thus, early detection of this *LAMB3* mutation provides a clinical opportunity for secondary prevention and reducing disease morbidity.

B. Prenatal diagnosis

Elucidation of specific disease associated gene mutations has had an enormous impact on the feasibility of prenatal diagnosis. In less than two decades, testing has progressed from mid-trimester fetal skin biopsies or protein analysis in a limited number of conditions to first-trimester chorionic villus sampling in a much broader range of genodermatoses. The first fetal skin biopsy diagnostic test for EB was reported 20 years ago (Rodeck *et al.*, 1980). Subsequently, apart from the life-threatening forms of EB, fetal skin sampling has been performed in pregnancies at risk for recurrence of EH, harlequin ichthyosis (HI), lamellar ichthyosis (LI), Sjogren-Larsson syndrome, subtypes of ectodermal dysplasia, and tyrosinase negative oculocutaneous albinism (Eady and McGrath, 1999). With the advances in molecular diagnostics, most of these disorders have now been tested for using DNA analysis at an earlier stage of the pregnancy. However, in some conditions, such as HI, the candidate gene remains to be identified. In addition, disorders such as JEB and LI show considerable genetic heterogeneity, and testing should be based on mutation detection alone rather than linkage analysis. Indeed, for EB, some procedural guidelines have been proposed for investigators undertaking DNA-based tests in pregnancies at risk for recurrence of dystrophic or junctional forms of EB (Christiano *et al.*, 1996, 1997). It is evident that in most instances it is important to have DNA samples from both parents and the previously affected individual to determine the pathogenic mutations. Other considerations such as the occurrence of *de novo* mutations, nonpaternity, uniparental disomy, and germline mosaicism can then all be addressed more fully and the suitability of the prenatal test can be determined.

Advances in *in vitro* fertilization protocols and embryo manipulation technology have further led to the feasibility of even earlier prenatal diagnosis through preimplantation genetic diagnosis (McGrath and Handyside, 1998). With this approach, couples known to be carrying pathogenic gene defects undergo the same process of ovary stimulation, egg retrieval, and *in vitro* fertilization as couples being treated for infertility. Fertilized embryos are cultured to cleavage stages, single cells are removed, or biopsied, for genetic analysis and, following diagnosis, unaffected embryos are selected for transfer back to the uterus. Preimplantation genetic diagnosis, therefore, represents a further clinically related advance and is applicable to several autosomal genodermatoses as well as X-linked disorders, such as incontinentia pigmenti.

C. Newer forms of therapy

One of the significant shortcomings associated with the identification of specific disease genes has been in the portrayal or interpretation of initial laboratory data as an indication that corrective gene therapy is just around the corner. In many ways, such a development remains a quantum leap rather than a small step. Nevertheless, significant progress is being made in the drive toward therapeutic gene manipulation, some of which may be applicable to a number of diseases or more focused to a specific gene. For example, recent studies have demonstrated the ability of aminoglycoside antibiotics to result in mistranslation and read-through of nonsense mutations (Barton-Davis *et al.*, 1999). At a cellular level, muscle cells bearing nonsense mutations on both alleles of the dystrophin gene were able to express some full-length protein after treatment with gentamicin. The significance of these findings to other inherited disorders, including a range of genodermatoses underscored by nonsense mutations, remains to be determined. Also relevant to several diseases is the observation that many disorders are complicated by excessive scarring, for example, the dystrophic forms of EB and some photogenodermatoses. Research based on antiscarring technology, such as the use of transforming growth factor beta-neutralizing antibodies or mannose-6-phosphate (which keeps this cytokine in its latent form) may have direct relevance to such patients (Shah *et al.*, 1992, 1999).

   As an organ for gene manipulation, the skin has already demonstrated considerable potential (Hengge *et al.*, 1999). Transduced fibroblasts or keratinocytes may deliver systemic proteins such as adenosine deaminase, human growth hormone, apolipoprotein E, β-glucuronidase, or α-galactosidase, or synthesize proteins locally in the skin, for example, laminin 5 or type XVII collagen in the treatment of junctional EB (Gagnoux-Palacios *et al.*, 1996; Dellambra *et al.*, 1998; Vailly *et al.*, 1998; Seitz *et al.*, 1999) and transglutaminase 1 in lamellar ichthyosis (Choate and Khavari, 1997). Gene replacement, either *in vivo* (direct injection) or *ex vivo* (grafting of transfected cultured keratinocytes), appears to be the appropriate strategy in gene therapy techniques for recessive disorders characterized by nonsense mutations and low/absent levels of the corresponding gene product. However, aside from gene replacement, other approaches are being explored, such as gene correction by homologous recombination, and for dominant disorders antisense and ribozyme technologies are all being pursued. Several difficult issues remain, including stem cell identification (Cotsarelis *et al.*, 1999), optimal vector design and delivery systems, and the need to sustain long-term *in vivo* gene expression (Deng *et al.*, 1997; Krueger *et al.*, 1999), but therapeutic gene manipulation remains a promising opportunity in attempting to improve management for a whole range of inherited skin disorders for which adequate treatments are currently lacking.

References

Aberdam, D., Galliano, M. F., Vailly, J., Pulkkinen, L., Bonifas, J., Christiano, A. M., Tryggvason, K., Uitto, J., Epstein, E. H. Jr., Ortonne, J. P., and Meneguzzi, G. (1994). Herlitz's junctional epidermolysis bullosa is linked to mutations in the gene (LAMC2) for the gamma 2 subunit of nicein/kalinin (LAMININ-5). *Nat. Genet.* **6**, 299–304.

Ahmad, W., Faiqaz ul Haque, M., Brancolini, V., Tsou, H. C., ul Haque, S., Lam, H., Aita, V. M., Owen, J., de Blaquiere, M., Frank, J., Cserhalmi-Friedman, P. B., Leask, A., McGrath, J. A., Peacocke, M., Ahmad, A., Ott, J., and Christiano, A. M. (1998a). Alopecia universalis associated with a mutation in the human hairless gene. *Science* **279**, 720–724.

Ahmad, W., Irvine, A. D., Lam, H, Buckley, C., Bingham, E. A., Panteleyev, A. A., Ahmad, M., McGrath, J. A., and Christiano, A. M. (1998b). A mis-sense mutation in the zinc-finger doman of the human hairless gene underlies congenital atrichia in a family of Irish travellers. *Am. J. Hum. Genet.* **63**, 984–991.

Ahmad, W., Zlotogorski, A., Panteleyev, A. A., Lam, H., Ahmad, M., ul Haque, M. F., Abdallah, H. M., Dragan, L., and Christiano, A. M. (1999). Genomic organization of the human hairless gene (HR) and identification of a mutation underlying congenital atrichia in an Arab Palestinian family. *Genomics* **56**, 141–148.

Armstrong, D. K., Hutchinson, T. H., Walsh, M. Y., and McMillan, J. C. (1997). Autosomal recessive inheritance of erythrokeratoderma variabilis. *Pediatr Dermatol* **14**, 355–358.

Armstrong, D. K., McKenna, K. E., Purkis, P. E., Green, K. J., Eady, R. A., Leigh, I. M., and Hughes, A. E. (1999). Haploinsufficiency of desmoplakin causes a striate subtype of palmoplantar keratoderma. *Hum. Mol. Genet.* **8**, 143–148.

Barton-Davis, E. R., Cordier, L., Shoturma, D. I., Leland, S. E., and Sweeney, H. L. (1999). Aminoglycoside antibiotics restore dystrophin function to skeletal muscles of mdx mice. *J. Clin. Invest.* **104**, 375–381.

Basarab, T., Smith, F. J., Joliffe, V. M., McLean, W. H., Neill, S., Rustin, M. H., and Eady, R. A. (1999). Ichthyosis bullosa of Siemens: Report of a family with evidence of a keratin 2e mutation, and a review of the literature. *Br. J. Dermatol.* **140**, 689–695.

Bonifas, J. M., Rothman, A. L., and Epstein, E. H. (1991). Epidermolysis bullosa simplex: Evidence in two families for keratin gene abnormalities. *Science* **254**, 1202–1205.

Borradori, L., and Sonnenberg, A. (1999). Structure and function of hemidesmosomes: More than simple adhesion complexes. *J. Invest. Dermatol.* **112**, 411–418.

Bowden, P. E., Haley, J. L., Kansky, A., Rothnagel, J. A., Jones, D. O., and Turner, R. J. (1995). Mutations of a type II keratin gene (K6a) in pachyonychia congenita. *Nat. Genet.* **10**, 363–365.

Bruckner-Tuderman, L. (1999). Hereditary skin diseases of anchoring fibrils. *J. Dermatol. Sci.* **20**, 122–133.

Choate, K. A., and Khavari, P. A. (1997). Direct cutaneous gene delivery in a human genetic skin disease. *Hum. Gene Therapy* **8**, 1659–1665.

Christiano, A. M., Greenspan, D. S., Hoffman, G. G., Zhang, X., Tamai, Y., Lin, A. N., Dietz, H. C., Hovnanian, A., and Uitto, J. (1993). A missense mutation in type VII collagen in two affected siblings with recessive dystrophic epidermolysis bullosa. *Nat. Genet.* **4**, 62–66.

Christiano, A. M., LaForgia, S., Paller, A. S., McGuire, J., Shimizu, H., and Uitto, J. (1996). Prenatal diagnosis for recessive dystrophic epidermolysis bullosa in 10 families by mutation and haplotype analysis in the type VII collagen gene (COL7A1). Mol. Med. 2, 59–76.

Christiano, A. M., Pulkkinen, L., McGrath, J. A., and Uitto, J. (1997). Mutation-based prenatal diagnosis of Herlitz junctional epidermolysis bullosa. Prenat. Diagn. 17, 343–354.

Cichon, S., Anker, M., Vogt, I. R., Rohleder, H., Putzstuck, M., Hillmer, A., Farooq, S. A., Al-Dhafri, K. S., Ahmad, M., Haque, S., Rietschel, M., Propping, P., Kruse, R., and Nothen, M. M. (1998). Cloning, genomic organisation, alternative transcripts and mutational analysis of the gene responsible for autosomal recessive universal congenital alopecia. Hum. Mol. Genet. 7, 1671–1679.

Cotsarelis, G., Kaur, P., Dhouailly, D., Hengge, U., and Bickenbach, J. (1999). Epithelial stem cells in the skin: Definition, markers, localization and functions. Exp. Dermatol. 8, 80–88.

Coulombe, P. A., Huwon, M. E., Letai, A., Hebert, A., Paller, A. S., and Fuchs, E. (1991). Point mutations in human keratin 14 genes of epidermolysis bullosa simplex patients: Genetic and functional analyses. Cell 66, 1201–1311.

Covello, S. P., Smith, F. J. D., Sillevis Smitt, J. H., Paller, A. S., Munro, C. S., Jonkman, M. F., Uitto, J., and McLean, W. H. (1998). Keratin 17 mutations cause either steatocystocytoma multiplex or pachyonychia congenita type 2. Br. J. Dermatol. 139, 475–480.

Dellambra, E., Vailly, J., Pellegrini, G., Bondanza, S., Golisano, O., Mackchia, C., Zambruno, G., Meneguzzi, G., and De Luca, M. (1998). Corrective transduction of human epidermal stem cells in laminin-5 deficient junctional epidermolysis bullosa. Hum. Gene Therapy 99, 1359–1379.

Deng, H., Lin, Q., and Khavari, P. A. (1997). Sustainable cutaneous gene delivery. Nat. Biotechnol. 15, 1388–1391.

de Viragh, P. A., Scharer, L., Bundman, D., and Roop, D. R. (1997). Loricrin deficient mice: upregulation of other cell envelope precursors rescues the neonatal defect but fails to restore epidermal barrier function. J. Invest. Dermatol. 108, 555 (abstract).

DiGiovanna, J. J., and Bale, S. J. (1994). Clinical heterogeneity in epidermolytic hyperkeratosis. Arch. Dermatol. 130, 1026–1035.

Dunnill, M. G., McGrath, J. A., Richards, A. J., Christiano, A. M., Uitto, J., Pope, F. M., and Eady, R. A. (1996). Clinicopathological correlations of compound heterozygous COL7A1 mutations in recessive dystrophic epidermolysis bullosa. J. Invest. Dermatol. 107, 171–177.

Eady, R. A., and McGrath, J. A. (1999). Genodermatoses. In "Fetal Medicine" (C. H. Rodeck and M. J. Whittle, eds.), Churchill Livingstone, London. pp. 545–552.

Fine, J. D., Eady, R. A. J., Bauer, E. A., Briggaman, R. A., Bruckner-Tuderman, L., Christiano, A. M., Heagerty, A. D., Jonkman, M. F., McGuire, J., Moshell, A., Shimizu, H., Tadini, G., and Uitto, J. (2000). Revised classification system for inherited epidermolysis bullosa: Report of the Second International Consensus Meeting on Diagnosis and Classification of Epidermolysis Bullosa. J. Am. Acad. Dermatol. 42, 1051–1066.

Fischer, J., Blanchet-Bardon, C., Prud'homme, J. F., Pavek, S., Steijlen, P. M., Dubertret, L., and Weissenbach, J. (1997). Mapping of Papillon-Lefevre syndrome to the chromosome 11q14 region. Eur. J. Hum. Genet. 51, 56–60.

Fuchs, E., Esteves, R. A., and Coulombe, P. A. (1992). Transgenic mice expressing a mutant K10 gene reveal the likely genetic basis for epidermolytic hyperkeratosis. Proc. Natl. Acad. Sci. (USA) 89, 6906–6910.

Gagnoux-Palacios, L., Vailly, J., Durand-Clement, M., Wagner, E., Ortonne, J. P., and Meneguzzi, G. (1996). Functional re-expression of laminin-5 in laminin-gamma2-deficient human keratinocytes modifies cell mophology, motility and adhesion. J. Biol. Chem. 271, 18437–18444.

Hammami-Hauasli, H., Schumann, H., Raghunath, M., Kilgus, O., Luthi, U., Luger, T., and Bruckner-Tuderman, L. (1998). Some, but not all, glycine substitution mutations in COL7A1 result in intracellular accumulation of collagen VII, loss of anchoring fibrils, and skin blistering. J. Biol. Chem. 273, 19228–19234.

Hashimoto, I., Kon, A., Tamai, K., and Uitto, J. (1999). Diagnostic dilemma of "sporadic" cases of dystrophic epidermolysis bullosa: A new dominant or mitis recessive mutation? *Exp. Dermatol.* **8,** 140–142.

Henegge, U. R., Taichman, L. B., Kaur, P., Rogers, G., Jensen, T. G., Goldsmith, L. A., Rees, J. L., and Christiano, A. M. (1999). How realistic is cutaneous gene therapy? *Exp. Dermatol.* **8,** 419–431.

Hennies, H. C., Kuster, W., Wiebe, V., Krebsova, A., and Reis, A. (1998). Genotype/phenotype correlation in autosomal recessive lamellar ichthyosis. *Am. J. Hum. Genet.* **62,** 1052–1061.

Hilal, L., Rochat, A., Duquesnoy, P., Blanchet-Bardon, C., Wechsler, J., Martin, N., Christiano, A. M., Barradon, Y., Uitto, J., Goossens, M., and Hovnanian, A. (1993). A homozygous insertion-deletion in the type VII collagen gene (COL7A1) in Hallopeau-Siemens dystrophic epidermolysis bullosa. *Nat. Genet.* **5,** 287–293.

Hovnanian, A., Duquesnoy, P., Amselem, S., Blanchet-Bardon, C., Lathrop, M, Dubertret, L., and Goossens, M. (1991). Exclusion of linkage between the collagenase gene and generalized recessive dystrophic epidermolysis bullosa phenotype. *J. Clin. Invest.* **88,** 1716–1721.

Hu, Z., Bonifas, J. M., Beech, J., Bench, G., Shigihara, T., Ogawa, H., Ikeda, S., Mauro, T., and Epstein, E. H. (2000). Mutations in *ATP2C1*, encoding a calcium pump, cause Hailey-Hailey disease. *Nat. Genet.* **24,** 61–65.

Huber, M., Rettler, I., Bernasconi, K., Frenk, E., Lavrijsen, S. P., Ponec, M., Bon, A., Lautenschlager, S., Schorderet, D. F., and Hohl, D. (1995). Mutations of keratinocyte transglutaminase in lamellar ichthyosis. *Science* **267,** 525–528.

Irvine, A. D., and McLean, W. H. (1999). Human keratin diseases: the increasing spectrum of disease and subtlety of the phenotype-genotype correlation. *Br. J. Dermatol.* **140,** 815–828.

Ishida-Yamamoto, A., McGrath, J. A., Chapman, S. J., Leigh, I. M., Lane, E. B., and Eady, R. A. (1991). Epidermolysis bullosa simplex (Dowling-Meara type) is a genetic disease characterized by an abnormal keratin filament network involving keratins K5 and K14. *J. Invest. Dermatol.* **97,** 959–968.

Ishida-Yamamoto, A., McGrath, J. A., Judge, M. R., Leigh, I. M., Lane, E. B., and Eady, R. A. (1992). Selective involvement of keratins K1 and K10 in the cytoskeletal abnormality of epidermolytic hyperkeratosis (bullous congenital ichthyosiform erythroderma). *J. Invest. Dermatol.* **99,** 19–26.

Ishida-Yamamoto, A., McGrath, J. A., Lam, H., Iizuka, H., Friedman, R. A., and Christiano, A. M. (1997). The molecular pathology of progressive symmetric erythrokeratoderma: A frameshift mutation in the loricrin gene and perturbations in the cornified cell envelope. *Am. J. Hum. Genet.* **61,** 581–589.

Ishida-Yamamoto, A., Takahashi, H., and Iizuka, H. (1998). Loricrin and human skin diseases: Molecular basis of loricrin keratodermas. *Histol. Histopathol.* **13,** 819–826.

Ishida-Yamamoto, A., and Iizuka, H. (1998). Structural organization of cornified cell envelopes and alterations in inherited skin disorders. *Exp. Dermatol.* **7,** 1–10.

Kelsell, D. P., Dunlop, J., Stevens, H. P., Lench, N. J., Liang, J. N., Parry, G., Mueller, R. F., and Leigh, I. M. (1997). Connexin 26 mutations in hereditary non-syndromic sensorineural deafness. *Nature* **387,** 80–83.

Kelsell, D. P., and Stevens, H. P. (1999). The palmoplantar keratodermas: Much more than palms and soles. *Mol. Med. Tod.* **5,** 107–113.

Kivirikko, S., McGrath, J. A., Baudoin, C., Aberdam, D., Ciatti, S., Dunnill, M. G. S., McMillan, J. R., Eady, R. A. J., Ortonne, J. P., Meneguzzi, G., Uitto, J., and Christiano, A. M. (1995). A homozygous nonsense mutation in the alpha 3 chain gene of laminin 5 (LAMA3) in lethal (Herlitz) junctional epidermolysis bullosa. *Hum. Mol. Genet.* **4,** 959–962.

Korge, B. P., Ishida-Yamamoto, A., Punter, C., Dopping-Hepenstal, P. J., Iizuka, H., Stephenson, A., Eady, R. A., and Munro, C. S. (1997). Loricrin mutation in Vohwinkel's keratoderma is unique to the variant with ichthyosis. *J. Invest. Dermatol.* **109,** 604–610.

Korge, B. P., Hamm, H., Jury, C. S., Traupe, H., Irvine, A. D., Healy, E., Birch-MacHin, M., Rees, J. L., Messenger, A. G., Holmes, S. C., Parry, D. A., and Munro, C. S. (1999). Identification of novel mutations in basic hair keratins hHb1 and hHb6 in monilethrix: implications for protein structure and clinical phenotype. *J. Invest. Dermatol.* **113,** 607–612.

Kowalczyk, A. P., Bornslaeger, E. A., Norvell, S. M., Palka, H. L., and Green, K. J. (1999). Desmosomes: Intercellular adhesive junctions specialized for attachment of intermediate filaments. *Int. Rev. Cytol.* **185**, 237–302.

Krueger, G. G., Morgan, J. R., and Petersen, M. J. (1999). Biologic aspects of expression of stably integrated transgenes in cells of the skin in vitro and in vivo. *Proc. Assoc. Am. Phys.* **111**, 198–205.

Kruse, R., Cichon, S., Anke, M., Hillmer, A. L., Barros-Nunez, P., Cantu, J. M., Leal, E., Weinlich, G., Schmuth, M., Fritsch, P., Ruzicka, T., Propping, P., and Nothen, M. M. (1999). Novel hairless mutations in two kindreds with autosomal recessive papular atrichia. *J. Invest. Dermatol.* **113**, 954–959.

Lane, E. B., Rugg, E. L., Navsaria, H., Leigh, I. M., Heagerty, A. H., Ishida-Yamamoto, A., and Eady, R. A. (1992). A mutation in the conserved helix termination peptide of keratin 5 in hereditary skin blistering. *Nature* **356**, 244–246.

Lazarides, E. (1980). Intermediate filaments as mechanical integrators of cellular space. *Annu. Rev. Biochem.* **51**, 219–250.

Lloyd, C., Yu, Q. C., Cheng J., Turksen, K., Degenstein, L., Hutton, E., and Fuchs, E. (1995). The basal keratin network of stratified squamous epithelia defining K15 function in the absence of K14. *J. Cell Biol.* **129**, 1329–1344.

Maestrini, E., Monaco, A. P., McGrath, J. A., Ishida-Yamamoto, A., Camisa, C., Hovnanian, A., Weeks, D. E., Lathrop, M., Uitto, J., and Christiano, A. M. (1996). A molecular defect in loricrin, the major component of the cornified cell envelope, underlies Vohwinkel's Syndrome. *Nat. Genet.* **13**, 70–77.

Maestrini, E., Korge, B. P., Ocana-Sierra, J., Calzolari, E., Cambiaghi, S., Scudder, P. M., Hovnanian, A., Monaco, A. P., and Munro, C. S. (1999). A missense mutation in connexin 26, D66H, causes mutilating keratoderma with sensorineural deafness (Vohwinkel's Syndrome) in three unrelated familes. *Hum. Mol. Genet.* **8**, 1237–1243.

Marinkovich, M. P. (1999). Update on inherited bullous dermatoses. *Dermatol. Clin.* **17**, 473–485.

McGrath, J. A., Ishida-Yamamoto, A., O'Grady, A., Leigh, I. M., and Eady, R. A. (1993). Structural variations in anchoring fibrils in dystrophic epidermolysis bullosa: Correlation with type VII collagen expression. *J. Invest. Dermatol.* **100**, 366–372.

McGrath, J. A., Gatalica, B., Christiano, A. M., Li, K., Owaribe, K., McMillan, J. R., Eady, R. A. J., and Uitto, J. (1995). Mutations in the 180-kDa bullous pemphigoid antigen (BPAG2), a hemidesmosomal transmembrane collagen (COL17A1), in generalized atrophic benign epidermolysis bullosa. *Nat. Genet.* **11**, 83–86.

McGrath, J. A., Christiano, A. M., Pulkkinen, L., Eady, R. A., and Uitto, J. (1996). Compound heterozygosity for nonsense and missense mutations in the LAMB3 gene in non-lethal junctional epidermolysis bullosa. *J. Invest. Dermatol.* **106**, 1157–1159.

McGrath, J. A., McMillan, J. R., Shemanko, C. S., Rundswick, S. K., Leigh, I. M., Lane, E. B., Garrod, D. R., and Eady, R. A. (1997). Mutations in plakophilin 1 result in ectodermal dysplasia/skin fragility syndrome. *Nat. Genet.* **17**, 240–244.

McGrath, J. A., and Handyside, A. H. (1998). Preimplantation genetic diagnosis of severe inherited skin diseases. *Exp. Dermatol.* **7**, 65–72.

McGrath, J. A., Hoeger, P. H., Christiano, A. M., McMillan, J. R., Mellerio, J. E., Ashton, G. H., Leigh, I. M., Lake, B. D., Harper, J. I., and Eady, R. A. (1999a). Skin fragility and hypohidrotic ectodermal dysplasia resulting from ablation of plakophilin 1. *Br. J. Dermatol.* **140**, 297–307.

McGrath, J. A., Ashton, G. H., Mellerio, J. E., Salas-Alanis, J. C., Swensson, O., McMillan, J. R., and Eady, R. A. (1999b). Moderation of phenotypic severity in dystrophic and junctional forms of epidermolysis bullosa through in-frame skipping of exons containing nonsense or frameshift mutations. *J. Invest. Dermatol.* **113**, 314–321.

McLean, W. H. I., Rugg, E. L., Lunny, D. P., Morley, S. M., Lane, E. B., Swensson, O., Dopping-Hepenstal, P. J., Griffiths, W. A., Eady, R. A., Higgins, C., Leigh, I. M., and Lane, E. B. (1995). Keratin-16 and keratin-17 mutations cause pachyonychia congenita. *Nat. Genet.* **9**, 273–278.

McLean, W. H. I., Pulkkinen, L., Smith, F. J., Rugg, E. L., Bullrich, F., Burgeson, R. E., Amano, S.,

Hudson, D. L., Owaribe, K., McGrath, J. A., McMillan, J. R., Eady, R. A., Leigh, I. M., Christiano, A. M., and Uitto, J. (1996). Loss of plectin causes epidermolysis bullosa with muscular dystrophy: cDNA cloning and genomic organization. *Genes Dev.* **10,** 1724–1735.

Mellerio, J. E., Smith, F. J., McMillan, J. R., McLean, W. H., McGrath, J. A., Morrison, G. A., Tierney, P., Albert, D. M., Wiche, G., Leigh. I. M., Geddes, J.. F., Lane, E. B., Uitto, J., and Eady, R. A. (1997). Recessive epidermolysis bullosa simplex associated with plectin mutations: Infantile respiratory complications in two unrelated cases. *Br. J. Dermatol.* **137,** 898–906.

Mellerio, J. E., Pulkkinen, L., McMillan, J. R., Lake, B. D., Horn, H. M., Tidman, M. J., Harper, J. I., McGrath, J. A., Uitto, J., and Eady, R. A. (1998a). Pyloric atresia-junctional epidermolysis bullosa syndrome: Mutations in the integrin β4 gene (ITGB4) in two unrelated patients with mild disease. *Br. J. Dermatol.* **139,** 862–871.

Mellerio, J. E., Eady, R. A., Atherton, D. J., Lake, B. D., and McGrath, J. A. (1998b). E210K mutation in the gene encoding the α3 chain of laminin-5 (LAMB3) is predictive of a phenotype of generalised atrophic benign epidermolysis bullosa. *Br. J. Dermatol.* **139,** 325–331.

Mellerio, J. E. (1999). Molecular pathology of the cutaneous basement membrane zone. *Clin. Exp. Dermatol.* **24,** 25–32.

North, A. J., Bardsley, W. G., Hyam, J., Bornslaeger, E. A., Cordingley, H. C., Trinnaman, B., Hatzfeld, M., Green, K. J., Magee, A. I., and Garrod, D. R. (1999). Molecular map of the desmosomal plaque. *J. Cell Sci.* **112,** 4325–4326.

Paller, A. S., Syder, A. J., Chan, Y. M., Yu, Q. C., Hutton, E., Tadini, G., and Fuchs, E. (1994). Genetic and clinical mosaicism in a type of epidermal nevus. *N. Engl. J. Med.* **331,** 1408–1415.

Papadavid, E., Koumantaki, E., and Dawber, R. P. (1998). Erythrokeratoderma variabilis: Case report and review of the literature. *J. Eur. Acad. Dermatol. Venereol.* **11,** 180–183.

Parmentier, L., Lakhdar, H., Blanchet-Bardon, C., Marchand, S., Dubertret, L., and Weissenbach, J. (1996). Mapping of a second locus for lamellar ichthyosis to chromosome 2q33-35. *Hum. Mol. Genet.* **5,** 555–559.

Parmentier, L., Clepet, C., Boughdene-Stambouli, O., Lakhdar, H., Blanchet-Bardon, C., Dubertret, L., Wunderle, E., Pulcini, F., Fizamnes, C., and Weissenbach, J. (1999). Lamellar ichthyosis, further narrowing, physical and expression mapping of the chromosome 2 candidate locus. *Eur. J. Hum. Genet.* **7,** 77–87.

Pulkkinen, L., Christiano, A. M., Airenne, T., Haakana, H., Tryggvason, K., and Uitto, J. (1994a). Mutations in the gamma 2 chain gene (LAMC2) of kalinin/laminin 5 in the junctional forms of epidermolysis bullosa. *Nat. Genet.* **6,** 293–297.

Pulkkinen, L., Christiano, A. M., Gerecke, D., Wagman, D. W., Burgeson, R. E., Pittelkow, M. R., and Uitto, J. (1994b). A homozygous nonsense mutation in the beta 3 chain gene of laminin 5 (LAMB3) in Herlitz junctional epidermolysis bullosa. *Genomics* **24,** 352–360.

Pulkkinen, L., Kimonis, V. E., Xu, Y., Spanou, E. N., McLean, W. H. I., and Uitto, J. (1997). Homozygous alpha6 integrin mutation in junctional epidermolysis bullosa with congenital duodenal atresia. *Hum. Mol. Genet.* **6,** 669–674.

Reis, A., Hennies, H. C., Langbein, L., Digweed, M., Mischke, D., Drechsler, M., Schrock, E., Royer-Pokara, B., Franke, W. W., and Sperling, K. (1994). Keratin 9 gene mutations in epidermolytic palmoplantar hyperkeratosis (EPPK). *Nat. Genet.* **6,** 174–179.

Richard, G., Smith, L. E., Bailey, R. A., Itin, P., Hohl, D., Epstein, E. H. Jr., DiGiovanna, J. J., Compton, J. G., and Bale, S. J. (1998). Mutations in the human connexin gene *GJB3* cause erythrokerato-dermia variabilis. *Nat. Genet.* **20,** 366–369.

Rickman, L., Simrak, D., Stevens, H. P., Hunt, D. M., King, I. A., Bryant, S. P., Eady, R. A., Leigh, I. M., Arnemann, J., Magee, A. I., Kelsell, D. P., and Buxton, R. S. (1999). N-terminal deletion in a desmosomal cadherin causes the autosomal dominant skin disease striate palmoplantar keratoderma. *Hum. Mol. Genet.* **8,** 971–976.

Rodeck, C. H., Eady, R. A., and Gosden, C. M. (1980). Prenatal diagnosis of epidermolysis bullosa letalis. *Lancet* **i,** 949–952.

Rothnagel, J. A., Dominey, A. M., Dempsey, L. D., Longley, M. A., Greenhalgh, D. A., Gague, T. A., Huber, M., Frenk, E., Hohl, D., and Roop, D. R. (1992). Mutations in the rod domains of keratin 1 and 10 in epidermolytic hyperkeratosis. *Science* **257**, 1128–1130.

Rouan, F., Pulkkinen, L., Meneguzzi, G., LaForgia, S., Hyde, P., Kim, D. S., Richard, G., and Uitto, J. (2000). Epidermolysis bullosa: Novel and *de novo* premature termination codon and deletion mutations in the plectin gene predict late-onset muscular dystrophy. *J. Invest. Dermatol.* **114**, 381–387.

Rugg, E. L., Morley, S. M., Smith, F. J. D., Boxer, M., Tidman, M. J., Navsaria, H., Leigh, I. M., and Lane, E. B. (1993). Missing links: Weber-Cockayne keratin mutations implicate the central L12 linker domain in effective cytoskeleton function. *Nat. Genet.* **5**, 294–300.

Rugg, E. L., McLean, W. H. I., Lane, E. B., Pitera, R., McMillan, J. R., Dopping-Hepenstal, P. J., Navsaria, H., Leigh, I. M., and Eady, R. A. (1994). A functional "knockout" for human keratin 14. *Genes Dev.* **8**, 2563–2573.

Russell, L. J., DiGiovanna, J. J., Rogers, G. R., Steinert, P. M., Hashem, N., Compton, J. G., and Bale, S. J. (1995). Mutations in the gene for transglutaminase 1 in autosomal recessive lamellar ichthyosis. *Nat. Genet.* **9**, 279–283.

Ruzzi, L., Gagnoux-Palacios, L., Pinola, M., Belli, S., Meneguzzi, G., D'Alessio, M., and Zambruno, G. (1997). A homozygous mutation in the integrin alpha6 gene in junctional epidermoloysis bullosa with pyloric atresia. *J. Clin. Invest.* **99**, 2826–2831.

Sakuntabhai, A., Ruiz-Perez, V., Carter, S., Jacobsen, N., Burge, S., Smith, M., Munro, C. S., O'Donovan, M., Craddock, N., Kucherlapati, R., Rees, J. L., Owen, N., Lathrop, G. M., Monaco, A. P., Strachan, T., and Hovnanian, A. (1999). Mutations in ATP2A2, encoding a Ca^{2+} pump, cause Darier disease. *Nat. Genet.* **21**, 271–277.

Seitz, C. S., Giudice, G. J., Balding, S. D., Marinkovich, M. P., and Khavari, P. A. (1999). BP180 gene delivery in junctional epidermolysis bullosa. *Gene. Therapy* **6**, 42–47.

Shah, M., Foreman, D. M., and Ferguson, M. W. (1992). Control of scarring in adult wounds by neutralising antibody to transforming growth factor beta. *Lancet* **339**, 213–214.

Shah, M., Revis, D., Herrick, S., Baillie, R., Thorgeirson, S., Ferguson, M. W., and Roberts, A. (1999). Role of elevated plasma transforming growth factor beta 1 levels in wound healing. *Am. J. Pathol.* **154**, 1115–1124.

Shimizu, H., Hammami-Hauasli, N., Hatta, N., Nishikawa, T., and Bruckner-Tuderman, L. (1999). Compound heterozygosity for silent and dominant glycine substitution mutations in COL7A1 leads to a marked transient intracytoplasmic retention of procollagen VII and a moderately severe dystrophic epidermolysis bullosa phenotype. *J. Invest. Dermatol.* **113**, 419–421.

Smith, F. J. D., Eady, R. A. J., Leigh, I. M., McMillan, J. R., Rugg, E. L., Kelsell, D. P., Bryant, S. P., Spurr, N. K., Geddes, J. F., Kirtschig, G., Milana, G., de Bono, A. G., Owaribe, K., Wiche, G., Pulkkinen, L., Uitto, J., McLean, W. H., and Lane, E. B. (1996). Plectin deficiency results in muscular dystrophy with epidermolysis bullosa. *Nat. Genet.* **13**, 450–457.

Smith, F. J. D., Jonkman, M. F., Van Goot, H., Coleman, C. M., Covello, S. P., Uitto, J., and McLean, W. H. (1998a). A mutation in human keratin K6b produces a phenocopy of the K17 disorder pachyonychia congenita type 2. *Hum. Mol. Genet.* **7**, 1143–1148.

Smith, F. J. D., Steijlen, P. M., McKenna, K., Healey, E., Rees, J. L., Fisher, M. P., Bonifas, J. M., Epstein, E. H. Jr, McKusick, V. A., Tan, E., Uitto, J., and McLean, W. H. (1998b). Cloning of multiple K16 genes and genotype-phenotype correlation in pachyonychia congenita type 1 and focal palmoplantar keratoderma. *J. Invest. Dermatol.* **110**, 502 (abstract).

Suga, Y., Ogawa, H., Bundman, D., and Roop, D. R. (1999). Transgenic mice expressing a mutant form of loricrin exhibit an erythrokeratodermia similar to Vohwinkel syndrome. *J. Invest. Dermatol.* **112**, 551 (abstract).

Suga, Y., Arin, M. J., Scott, G., Goldsmith, L. A., Magro, C. M., Baden, L. A., Baden, H. P., and Roop, D. R. (2000). Hotspot mutations in keratin 2e suggest a correlation between genotype and phenotype in patients with ichthyosis bullosa of Siemens. *Exp. Dermatol.* **9**, 11–15.

Sybert, V. P., Francis, J. S., Corden, L. D., Smith, L. T., Weaver, M., Stephens, K., and McLean, W. H. (1999). Cyclic ichthyosis with epidermolytic hyperkeratosis: A phenotype conferred by mutations in the 2B domain of keratin K1. *Am. J. Hum. Genet.* **64,** 732–738.

Terracina, M., Posteraro, P., Schubert, M., Sonego, G., Atzori, F., and Zambruno, G. (1998). Compound heterozygosity for a recessive glycine substitution and a splice site mutation in the COL7A1 gene causes an unusually mild form of localized recessive dystrophic epidermolysis bullosa. *J. Invest. Dermatol.* **111,** 744–750.

Toomes, C., James, J., Wood, A. J., Wu, C. L., McCormick, D., Leach, N., Hewitt, C., Moynihan, L., Roberts, E., Woods, C. G., Markham, A., Wong, M., Widmer, R., Ghaffer, K. A., Pemberton, M., Hussein, I. R., Temtamy, S. A., Davies, R., Read, A. P., Sloan, P., Dixon, M. J., and Thakker, N. S. (1999). Loss-of-function mutations in the cathepsin C gene result in periodontal disease and palmoplantar keratosis. *Nat. Genet.* **23,** 421–424.

Uitto, J., Pulkkinen, L., and McLean, W. H. (1997). Epidermolysis bullosa: A spectrum of clinical phenotypes explained by molecular heterogeneity. *Mol. Med. Today* **3,** 457–465.

Uttam, J., Hutton, E., Coulombe, P. A., Anton-Lamprechy, I., Yu, Q. C., Gedde- Dahl, T. J. Jr., Fine, J. D., and Fuchs, E. (1996). The genetic basis of epidermolysis bullosa with mottled pigmentation. *Proc. Natl. Acad. Sci. (USA)* **93,** 9079–9084.

Vailly, J., Gagnoux-Palacios, L., Dell'Ambra, E., Romero, C., Pinola, M., Zambruno, G., De Luca, M., Ortonne, J. P., and Meneguzzi, G. (1998). Corrective gene transfer of keratinocytes from patients with junctional epidermolysis bullosa restores assembley of hemidesmosomes in reconstructed epithelia. *Gene. Therapy* **5,** 1322–1332.

van Steesel, M., Smith, F. J., Steijlen, P. M., Kluijt, I., Stevens, H. P., Messenger, A., Kremer, H., Dunnill, M. G., Kennedy, C., Munro, C. S., Doherty, V. R., McGrath, J. A., Covello, S. E., Coleman, C. M., Uitto, J., and McLean, W. H. (1999). The gene for hypotrichosis of Marie-Unna (MU) maps between D8S258 and D8S298: Exclusion of the *hr* gene by cDNA and genomic sequencing. *Am. J. Hum. Genet.* **65,** 413–419.

Vassar, R., Coulombe, P. A., Degenstein, L., Albers, K., and Fuchs, E. (1991). Mutant keratin expression in transgenic mice causes marked abnormalities resembling a human genetic skin disease. *Cell* **64,** 365–380.

Vidal, F., Aberdam, D., Miquel, C., Christiano, A. M., Pulkkinen, L., Uitto, J., Ortonne, J. P., and Meneguzzi, G. (1995). Integrin beta 4 mutations associated with junctional epidermolysis bullosa with pyloric atresia. *Nat. Genet.* **10,** 229–234.

Whittock, N. V., Ashton, G. H. S., Dopping-Hepenstal, P. J. C., Gratian, M. J., Keane, F. M., Eady, R. A. J., and McGrath, J. A. (1999a). Striate palmoplantar keratoderma resulting from desmoplakin haploinsufficiency. *J. Invest. Dermatol.* **113,** 940–946.

Whittock, N. V., Ashton, G. H. S., Mohammedi, R., Mellerio, J. E., Abbs, S., Mathew, C., Eady, R. A., and McGrath, J. A. (1999b). Comparative mutation detection of the type VII collagen gene (COL7A1) using the protein truncation test, fluorescent chemical cleavage of mismatch, and conformation sensitive gel electrophoresis. *J. Invest. Dermatol.* **113,** 673–686.

Wiche, G. (1998). Role of plectin in cytoskeleton organization and dynamics. *J. Cell Sci.* **111,** 2477–2486.

Winter, H., Rogers, M. A., Gebhardt, M., Wollina, U., Boxall, L., Chitayat, D., Babul-Hirji, R., Stevens, H. P., Zlotogorski, A., and Schweizer, J. (1997a). A new mutation in the type II hair cortex keratin hHb1 involved in the inherited hair disorder monilethrix. *Hum. Genet.* **101,** 765–769.

Winter, H., Rogers, M. A., Langbein, L., Stevens, H. P., Leigh, I. M., Labreze, C., Roul, S., Taieb, A., Krieg, T., and Schweizer, J. (1997b). Mutations in the hair cortex keratin hHb6 cause the inherited hair disease monilethrix. *Nat. Genet.* **16,** 372–374.

Yoneda, K., and Steinert, P. M. (1993). Overexpression of human loricrin in transgenic mice produces a normal phenotype. *Proc. Natl. Acad. Sci. (USA)* **90,** 10754–10758.

2

Molecular Genetics and Target Site Specificity of Retroviral Integration

Michelle L. Holmes-Son, Rupa S. Appa, and Samson A. Chow*
Department of Molecular and Medical Pharmacology
UCLA AIDS Institute and Molecular Biology Institute
UCLA School of Medicine
Los Angeles, California 90095

I. Retroviral Integration
   A. The integration reaction
   B. Essential components of the integration reaction
II. Target Site Selection during Retroviral Integration
   A. Factors affecting target site selection
   B. Role of integrase in target site selection
III. Site-Directed Integration
   A. Directed integration using integrase-fusion proteins *in vitro*
   B. Incorporating integrase-fusion proteins into infectious virions
   C. Future direction: site-specific retroviral vectors
IV. Concluding Remarks
   References

ABSTRACT

Integration is an essential step in the life cycle of retroviruses, resulting in the stable joining of the viral cDNA to the host cell chromosomes. While this critical process makes retroviruses an attractive vector for gene delivery, it also presents a potential hazard. The sites where integration occurs are nonspecific. Therefore,

*Corresponding author: Telephone: (310) 825-9600. Fax:(310) 825-6267. E-mail: schow@mednet. ucla.edu

Advances in Genetics, Vol. 43

it is possible that integration of retroviral DNA will affect host gene expression and disrupt normal cellular functions. The mechanism by which integration sites are chosen is not well understood, and is influenced by several factors, including DNA sequence and structure, DNA-binding proteins, DNA methylation, and transcription. Integrase, the viral enzyme responsible for catalyzing integration, also plays a key role in controlling the choice of target sites. The integrase domain responsible for target site selection has been mapped to the central core region. A better understanding of the interaction between the target-specifying motif of integrase and the target DNA may allow a means to manipulate integration into particular chromosomal sites. Another approach to directing integration is to fuse integrase with a sequence-specific DNA-binding protein, which results in a bias of integration *in vitro* into the recognition site of the fusion partner. Successful incorporation of the fusion protein into infectious virions and the identification of optimal proteins that can be fused to integrase will advance the development of site-specific vectors. Retroviruses are promising for the delivery of genes in experimental and therapeutic protocols. A better understanding of integration will aid in the design of safer and more effective gene transfer vectors.

© 2001 Academic Press.

I. RETROVIRAL INTEGRATION

Integration is one of the essential steps of the retroviral life cycle, of which many critical features remain poorly understood. This chapter will briefly cover what is known about the necessary components of the integration reaction and what occurs when they are brought together during retroviral infection. An emphasis is placed on factors influencing where integration takes place in the chromosomal DNA, mapping the domain of integrase that specifies those sites, and strategies for manipulating the site of integration. The following reviews are suggested readings (Brown, 1997; Chow, 1997; Goff, 1992; Katz and Skalka, 1994; Vink and Plasterk, 1993; Whitcomb and Hughes, 1992), as it is not possible to adequately cover all areas of retroviral integration in this chapter. Additional reviews will also be referenced throughout the chapter to provide further information on specific topics regarding integration when they are introduced.

A retrovirus initiates infection when it comes in contact with a cell that expresses the appropriate receptor and coreceptor on its surface. Envelope proteins on the surface of the viral particle recognize those specific receptors on the cell and interact with them to gain entry of the virus into the cytoplasm, by the process of membrane fusion. Once in the cytoplasm, the virus begins to reverse-transcribe its positive-strand RNA genome into double-stranded DNA, and then transports it to the chromosomes of the cell for integration. Integration of the viral cDNA results in the formation of a provirus, which then allows the retrovirus to exploit the host cellular transcriptional and translational machinery for the production of new viral genomic and messenger RNA, and proteins. The viral proteins and RNA genome are then assembled near the plasma membrane of the cell, bud off at the surface, and mature so that they can infect new cells to begin the next round of viral replication.

Despite this brief introduction to the viral life cycle, a few important points about integration are evident. First, integration is essential for retroviral replication. If the retrovirus is unable to integrate a copy of its genome into the cellular DNA, no new viral transcripts or proteins will be made, and infection is halted. Second, integration results in the permanent insertion of the viral genome into the chromosome of the target cell. Therefore, once this step in the viral life cycle is complete, the infected cell retains the potential to produce new viral particles for the duration of its lifetime. In addition, because the provirus is a part of the chromosomal DNA, as the cell divides, any new daughter cells will also inherit the ability to produce virions. Finally, by inserting the viral genome into host chromosomal DNA, the integration reaction is an inherently mutagenic process to the infected cell. Depending on the site of integration, insertional mutagenesis may disrupt normal cell functions.

A. The integration reaction

Integration of a cDNA copy of the retroviral genome into the host cellular DNA is a highly ordered three-step process, similar to reactions conducted by other members of the family of polynucleotidyl transferases (Mizuuchi, 1992; Rice *et al.*, 1996). In the first step, 3′-end processing, integrase cleaves off the two terminal bases at each 3′ end of the linear viral DNA, exposing a highly conserved CA dinucleotide on both strands (Figure 2.1A; Brown *et al.*, 1989; Fujiwara and Mizuuchi, 1988; Katzman *et al.*, 1989). Trimming of the viral ends by integrase may provide the virus with a means to maintain genomic consistency because reverse transcription can result in the incorporation of extra bases at the ends of the viral cDNA (Miller *et al.*, 1995). During 3′-end processing, integrase uses the hydroxyl group of a water molecule to attack the phosphodiester bond subsequent to the adenosine of the CA dinucleotide. The leaving group of this reaction is a free 3′ hydroxyl on that adenosine base (Engelman *et al.*, 1991; Gerton *et al.*, 1999). The next step of integration, 3′-end joining or strand transfer, uses this free 3′ hydroxyl to attack the phosphodiester backbone of the chromosomal DNA (Engelman *et al.*, 1991; Gerton *et al.*, 1999), joining the 3′ ends of the viral DNA to the chromosomes of the host cell (Bushman and Craigie, 1991; Bushman *et al.*, 1990; Craigie *et al.*, 1990; Katz *et al.*, 1990). This joining reaction occurs with a stereotypic spacing in the opposite strands of the host DNA. The spacing of 4, 5, or 6 base pairs is characteristic for each retroviral integrase. The chemistry of both 3′-end processing and 3′-end joining is a one-step transesterification reaction

A

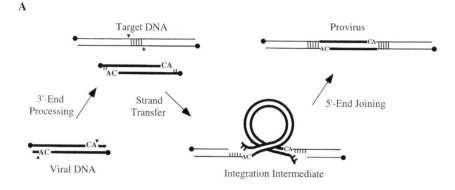

B

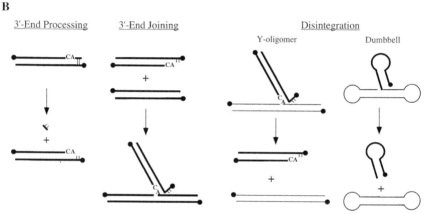

Figure 2.1. (A) Pathway of retroviral DNA integration. The integration reaction proceeds in three steps: 3′-end processing, 3′-end joining, and 5′-end joining. Filled triangles indicate sites of cleavage or 3′-end joining. (B) *In vitro* assays for integrase activity. The catalytic activity of integrase can be assayed using oligonucleotide substrates that mimic the end of the viral LTR, the target DNA, or the integration intermediate. 3′-End processing activity is assayed by the appearance of a product that is shortened by two nucleotides. In 3′-end joining, the processed substrate mediates a coupled cleavage–ligation reaction using another molecule as the target. The joining activity is assayed by the appearance of products that are longer in length than the input DNA. The lengths of the products are heterogeneous because the site of joining is nonspecific. Disintegration, a reversal of the 3′-end joining reaction, can be assayed by using either the Y-oligomer or the dumbbell substrate. Both disintegration substrates have a structure similar to the 3′-end joining product and the reaction can reverse itself to produce the separate viral and target DNA components. In both (A) and (B), thick lines represent viral DNA and thin lines represent target DNA. Filled circles denote the 5′ end of DNA. The conserved CA dinucleotide at the viral DNA end is indicated. Long vertical lines represent base pairs and short vertical lines represent unpaired nucleotides.

(Engelman *et al.*, 1991; Gerton *et al.*, 1999). The final step of integration, 5′-end joining, resolves the structure resulting from the processing and joining steps. This structure is referred to as the gapped intermediate because the 2 unpaired bases at both viral 5′ ends after 3′-end processing and the 4–6 unpaired bases in the genomic DNA flanking the viral sequence after 3′-end joining are still present (Figure 2.1A).

Although it is not completely understood whether host or viral processes are responsible for repair of this integration intermediate, recent evidence has indicated the potential for cellular DNA repair enzymes in this pathway (Daniel *et al.*, 1999). At a time point corresponding to the integration step, cells infected with a Rous sarcoma virus (RSV)-based vector displayed increased DNA-dependent protein kinase (DNA-PK) activity, an enzyme involved in double-stranded DNA break repair. Also, about 24 h after infection, with either an HIV-1– or RSV-based vector, while cells with a wild-type DNA-PK were largely unaffected, over 50% of cells containing a genetic defect in DNA-PK died of apoptosis (Daniel *et al.*, 1999). Despite this evidence, it remains possible that viral proteins, integrase and reverse transcriptase, could be responsible for this final step. Both proteins are present in the nucleoprotein complex that migrates into the nucleus to integrate the viral cDNA. Reverse transcriptase could fill the gaps on either side of the viral genome, while integrase, which has DNA splicing activity (Chow *et al.*, 1992), could remove the extra bases at the 5′ ends of the viral genome and seal the nicks at the viral–host DNA junction (Roe *et al.*, 1997). Because no direct evidence exists for either host or viral processes in this critical reaction, 5′-end joining remains poorly understood.

The result of integration of the retroviral DNA is provirus formation that is characterized by viral sequences that begin with a 5′-TG and terminate in CA-3′. The provirus is flanked by a short duplication of the genomic DNA, the length of which depends on the size of the stagger cut in the chromosome during the 3′-end joining reaction (Figure 2.1A; Brown *et al.*, 1989; Dhar *et al.*, 1980; Hughes *et al.*, 1981; Majors and Varmus, 1981; Shimotohno *et al.*, 1980).

The first two steps of integration, 3′-end processing and 3′-end joining, can be monitored in *in vitro* assays using purified integrase and short annealed oligonucleotides that mimic the ends of the viral DNA (Figure 2.1B; for review, see Chow, 1997). In the 3′-end processing reaction, the trimming of two nucleotides from the 3′ end of the viral DNA surrogate is measured. In the 3′-end joining reaction, the insertion of one of those ends into another DNA molecule is monitored.

B. Essential components of the integration reaction

For integration to occur, two factors are absolutely required: the viral enzyme integrase and sequences present at the ends of the viral DNA genome.

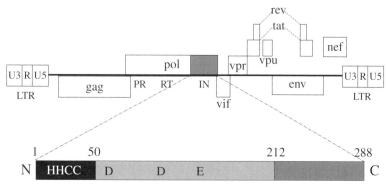

Figure 2.2. Schematic representation of the HIV-1 genome and primary structure of integrase. The integrase gene is enlarged to illustrate important features. The N-terminal domain (residues 1–50) contains the HHCC motif that binds zinc. The catalytic core domain (residues 50–212) harbors the active site DD(35)E residues of the enzyme. The C-terminus (residues 212–288) possesses nonspecific DNA-binding activity. The virus attachment sites are present at the terminal ends of the long terminal repeat (LTR).

1. Integrase

Integrase is encoded at the 3' end of the viral *pol* gene and has an approximate weight of 30–46 kDa, depending on the retrovirus. Although the structure of the full-length integrase has not yet been solved and many gaps remain in our understanding of the enzyme, extensive biochemical and genetic information is currently available (for reviews, see Asante-Appiah and Skalka, 1999; Esposito and Craigie, 1999; Wlodawar, 1999).

The structural organization of integrase consists of three independently folding domains, an N- and C-terminal domain, and a catalytic core, which were initially defined by limited proteolysis (Figure 2.2; Engelman and Craigie, 1992). The core of the protein is the most conserved region, and it contains the active site of the enzyme, characterized by a phylogenetically conserved D,D(35)E motif among retroviral integrases, bacterial insertional sequences, and some retrotransposons (Kulkosky *et al.*, 1992). Each of these residues is essential for catalysis, and mutating any of the three, even by conservative substitution, results in abolishment of integration (Drelich *et al.*, 1992; Engelman and Craigie, 1992; Gaur and Leavitt, 1998; Kulkosky *et al.*, 1992; Leavitt *et al.*, 1996). These residues are most likely involved in the coordination of a divalent metal ion, either Mg^{2+} or Mn^{2+}. Upon binding of the metal ion, HIV-1 integrase undergoes a reversible conformational change that activates the catalytic activity of the enzyme (Asante-Appiah *et al.*, 1998; Asante-Appiah and Skalka, 1997).

The core of integrase is also capable of recognizing features of both the viral and target DNA on specialized substrates referred to as disintegration substrates. The disintegration DNA substrate, which can be a Y-mer or dumbbell (Figure 2.1B), resembles the 3′-end joining product. It contains a viral DNA end that is joined at the 3′ adenosine residue to a double-stranded, nonspecific target DNA. Although the isolated core domain of integrase is unable to bind to linear DNA containing viral or nonspecific sequences independently, it is able to bind to the Y-mer disintegration substrates (Engelman et al., 1994). Activity assays using the disintegration substrate demonstrate that the purified core of HIV-1 integrase recognizes the highly conserved CA dinucleotide at the viral 3′ end (Gerton and Brown, 1997; Gerton et al., 1998). Furthermore, peptides corresponding to amino acid residues 49–69 and 139–152 of the HIV-1 integrase core domain photocrosslink to both the viral and target DNA features of the dumbbell disintegration substrate (Figure 2.1B; Heuer and Brown, 1997). Full-length integrase, when incubated with a short annealed oligonucleotide substrate that mimics the end of the viral DNA, is photocrosslinked to specific amino acid residues found in the core (Esposito and Craigie, 1998; Jenkins et al., 1997).

In addition to catalysis and substrate recognition, the core domain of integrase contains a determinant that specifies the site of integration along the chromosomal DNA (Katzman and Sudol, 1995, 1998; Shibagaki and Chow, 1997). This role of the core domain will be reviewed in greater detail in Section II.B.

The structures of the isolated core domains of human immunodeficiency virus type-1 (HIV-1) and avian sarcoma virus (ASV) integrases have been solved by X-ray crystallography (Bujacz et al., 1995, 1996a, 1996b; Dyda et al., 1995). The two contain a high degree of similarity with each other and a large family of polynucleotidyl transferases that include Escherichia coli RuvC of the Holliday junction resolvase, HIV-1 and E. coli RNase H, and the bacterial transposase, MuA (Dyda et al., 1995), indicating a clear relationship between structure and function.

The next conserved domain of retroviral integrases is in the N-terminus. It makes up approximately the first 50 amino acids and contains a highly conserved HHCC motif (Figure 2.2; Johnson et al., 1986). As expected, it is capable of binding zinc, and does so at a 1:1 ratio of integrase monomer to zinc ion for both HIV-1 (Burke et al., 1992; Lee et al., 1997; Zheng et al., 1996) and murine leukemia virus (MLV) integrases (Yang et al., 1999). Although the exact function of this domain is not yet clear, several important attributes have been assigned to it when it is bound to zinc ions. For HIV-1 and MLV integrases, stability is increased in the isolated domain from a disordered to a predominantly alpha-helical structure (Burke et al., 1992; Yang et al., 1999; Zheng et al., 1996). Multimerization of integrase is also promoted, and there is a stimulation of Mg^{2+}-dependent integration activity (Lee and Han, 1996; Lee *et al.*, 1997; Yang *et al.*, 1999; Zheng *et al.*, 1996).

The N-terminus of HIV-1 integrase alone does not bind DNA, though it may interact with viral or target DNA in the context of a full-length protein. Biochemical analyses of truncation mutants and chimeric integrases indicate that the N-terminal domain is involved in recognition of specific viral DNA ends (van den Ent *et al.*, 1999; Vincent *et al.*, 1993). However, a peptide consisting of the first 12 amino acids photocrosslinks with the target DNA portion of the dumbbell disintegration substrate (Figure 2.1B; Heuer and Brown, 1997). Further complicating our understanding of the function of the N-terminus of integrase is the different requirement of the N-terminus for *in vitro* activity of the various purified integrases. The N-terminal domain is dispensable for integration for ASV and RSV integrases (Bushman and Wang, 1994; Kahn *et al.*, 1991), but is absolutely essential for HIV-1, FIV, and MLV integrases (Bushman *et al.*, 1993; Jonsson *et al.*, 1996; Shibagaki *et al.*, 1997). In *in vivo* experiments, HIV-1 virions that contain mutations in either histidine of the HHCC motif of integrase indicate an additional role for the N-terminal domain. These viruses have a severe deficit in viral cDNA synthesis and, correspondingly, infectivity (Engelman *et al.*, 1995; Masuda *et al.*, 1995; Wu *et al.*, 1999). This implicates that the N-terminus of integrase is necessary for stimulating viral cDNA synthesis by interacting with the reverse transcription machinery.

Multidimensional heteronuclear nuclear magnetic resonance (NMR) structures have been solved for the N-terminal portion of HIV-1 (Cai *et al.*, 1997) and HIV-2 (Eijkelenboom *et al.*, 1997) integrases. The monomer folds of both N-terminal domains do not show similarity with any of the known zinc-finger structures, and instead are similar to those found in a number of helical DNA-binding proteins containing a helix–turn–helix (HTH) motif. However, unlike the DNA-binding proteins, the second helix of the HTH motif of integrase is not used for DNA recognition, but for dimerization.

The C-terminal domain of the retroviral integrases harbors no consistent amino acid motif, and overall is the least conserved. The NMR structure for the C-terminus of HIV-1 integrase has been solved and indicates that there is a src-homology 3-like fold in this domain (Eijkelenboom *et al.*, 1995; Lodi *et al.*, 1995). Although the significance of this information is not yet known, another important structural feature of the HIV-1 C-terminal domain was uncovered. At the dimer interface formed by the isolated domain in solution is a large saddle-shaped groove with an appropriate size and shape to fit a double-stranded DNA. This observation is consistent with biochemical experiments, such as UV crosslinking and Southwestern blotting, indicating that the C-terminal domain is capable of binding nonspecific DNA (Engelman *et al.*, 1994; Kahn *et al.*, 1991; Mumm and Grandgenett, 1991; Puras Lutzke *et al.*, 1994; Vink *et al.*, 1993; Woerner *et al.*, 1992; Woerner and Marcus-Sekura, 1993). In addition, mutation of one of several basic amino acids in the C-terminus results in a loss in ability to UV-crosslink this isolated region to DNA (Puras Lutzke and Plasterk, 1998; Puras Lutzke *et al.*, 1994). Because of its ability to bind nonspecific DNA, the function of the C-terminus has been postulated to bind target DNA. However, the C-terminus also interacts specifically with the subterminal sequence of viral DNA ends (Esposito and Craigie, 1998; Heuer and Brown, 1997), suggesting that the C-terminus may provide initial contact with viral DNA, and is mainly involved in stabilizing the interaction between integrase and viral DNA.

The active form of integrase is a multimer. Complementation experiments conducted with purified HIV-1 or MLV integrases revealed that monomers containing mutations or truncations in different domains, which are otherwise incapable of integration on their own, are functional when mixed in *in vitro* assays (Donzella *et al.*, 1998; Ellison *et al.*, 1995; Engelman *et al.*, 1993; van Gent *et al.*, 1993). Likewise, noninfectious HIV-1 particles harboring a mutation in the virally encoded integrase are functionally complemented *in trans* with an integrase containing an inactivating mutation in a different domain (Fletcher *et al.*, 1997). While it is accepted that integrase is active as a multimer, the exact number of monomers required for activity is unknown. Structural analysis of the core of HIV-1 and ASV integrases showed them to be dimers (Bujacz *et al.*, 1995, 1996a, 1996b; Dyda *et al.*, 1995), as did the N- and C-terminal domains of HIV-1 integrase (Eijkelenboom *et al.*, 1995, 1997; Lodi *et al.*, 1995). Gel filtration and sedimentation equilibrium analysis have indicated the presence of higher-order multimers of purified HIV-1 integrase in solution that include dimers, tetramers, and octamers (Jenkins *et al.*, 1996; Jones *et al.*, 1992; Lee *et al.*, 1997; Zheng *et al.*, 1996). HIV-1 virus particles were also probed for the multimeric state of integrase by fusing an antigenic epitope sequence to the protein. Western blot analysis of the lysed virions indicated that integrase is present predominantly as a dimer, although some higher-order species are present as well (Petit *et al.*, 1999).

During retroviral infection, integrase is not the only protein found on the viral cDNA prior to integration. Rather, a complex of viral and host cell proteins is formed following reverse transcription, and that then migrates to the chromosomes for integration. This complex is referred to as the preintegration complex (PIC). The HIV-1 PIC contains the viral proteins integrase, reverse transcriptase, matrix, Vpr, and perhaps nucleocapsid (Bukrinsky *et al.*, 1993; Farnet and Haseltine, 1990, 1991; Gallay *et al.*, 1995; Miller *et al.*, 1997). In addition, several host cell proteins are present as identified by salt-stripping PICs isolated from HIV-1- or MLV-infected cells. Each of these proteins plays an important role during retroviral integration. A barrier-to-autointegration factor (BAF) was initially found to be essential in preventing MLV PICs from integrating into their own viral cDNA, a suicidal process (Lee and Craigie, 1994, 1998; Li *et al.*, 1998), and has since been determined to stimulate integration of HIV-1 PICs (Chen and Engelman, 1998). HMG I(Y) has also been identified in the HIV-1

PIC, and is necessary to stimulate integration by salt-stripped complexes isolated from HIV-1-infected cells (Farnet and Bushman, 1997). HMG I(Y) also enhances integration by the MLV PIC (Li et al., 1998). The presence of the various host and viral factors in the nucleoprotein complex results in a large structure. The Stokes radius of the HIV-1 PIC has been estimated at 28 nm (Miller et al., 1997), and the sedimentation coefficient of MLV PICs at around 160S (Bowerman et al., 1989). Protection of the terminal 200–250 base pairs of viral cDNA also occurs because of protein binding in the PIC (Chen and Engelman, 1998; Chen et al., 1999; Wei et al., 1997). The formation of the nucleoprotein complex may protect viral cDNA ends from cellular nucleases, and bridge the two viral ends together to perform a concerted integration reaction (Miller et al., 1997).

2. Viral attachment sites

For integration to occur, specific sequences at the ends of the viral cDNA, referred to as attachment sites, must be present for integrase recognition. In vitro analysis using purified proteins and short oligonucleotides that mimic the ends of the viral DNA indicates that the most essential feature of the attachment sites is the highly conserved CA dinucleotide present at both 3′ termini of the viral cDNA, two bases internal to the ends. Deletion or mutation of either of these bases led to an ablatement of integration activity for all integrases tested (Bushman and Craigie, 1991; Ellison and Brown, 1994; Esposito and Craigie, 1998; Leavitt et al., 1992; Sherman et al., 1992; van den Ent et al., 1994; Vink et al., 1994). Similar mutational analyses that examined the catalytic activity (Balakrishnan and Jonsson, 1997; Bushman and Craigie, 1991; Esposito and Craigie, 1998; Katzman et al., 1989; LaFemina et al., 1991; Sherman et al., 1992; Vink et al., 1991) and ability of integrase to form a stable complex with viral DNA (Ellison and Brown, 1994; Vink et al., 1994), showed that additional bases, approximately 6–15 nucleotides from the viral DNA ends, are required for optimal in vitro integration activity. Mutations and deletion of terminal bases in viral clones have also been made to determine the length of DNA required at the attachment site for integration during retroviral infection. The terminal 12 bases have the most influence (Masuda et al., 1995, 1998), and the conserved CA dinucleotide at both viral 3′ ends is critical (Brown et al., 1999; Masuda et al., 1995, 1998; Reicin et al., 1995). Mutations of both CA dinucleotides completely abolish infectivity (Brown et al., 1999), and mutation at just one end decreases infectivity to about 20–40% of wild-type (Brown et al., 1999; Masuda et al., 1995, 1998).

II. TARGET SITE SELECTION DURING RETROVIRAL INTEGRATION

Retroviral integration occurs throughout the chromosomes. Initial studies analyzing integration sites in cells infected with ASV (Hughes et al., 1978), ALV

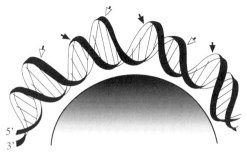

Figure 2.3. Distortion of B-DNA presents desirable sites for retroviral integration. Integration of retroviral DNA is preferred into sites that are bent due to intrinsic structure or protein binding. As depicted, the binding of DNA to certain proteins (shaded semicircle) creates DNA bends and makes the outer face accessible. The phosphodiester bonds, indicated by arrows, flanking the widened major groove are favorable sites for integration. Each pair of arrows (open and closed) represents an integration event. (Modified from Pryciak and Varmus, 1992.)

(Withers-Ward *et al.*, 1994), and RSV (Shih *et al.*, 1988) revealed that most regions of the genomic DNA are accessible to the PIC. Despite this ability to integrate the viral genome into virtually any site in the chromosomal DNA, integration does not occur randomly. Rather, using Southern blot and PCR-based techniques, it was determined that the same region, or exact nucleotide sequence in the genome, can be utilized at a frequency several hundred-fold greater than chance, lending to the idea that there are hot spots and cold spots for integration (Withers-Ward *et al.*, 1994). Therefore, integration of retroviral DNA into target DNA is nonspecific, but not a random process.

Since integration is essential for productive viral infection, the site of integration may be significant to the fate of both the virus and the host cell. The mechanism that determines target site selection is not well understood, but is probably affected by multiple factors. As detailed below, the salient feature that links the diverse factors in integration site selection is distortion of B-DNA, which makes the phosphodiester backbone more accessible and favorable for nucleophilic attack by the viral DNA substrate (Figure 2.3).

A. Factors affecting target site selection

1. Chromatin assembly and DNA bending

Initial experiments investigating factors affecting retroviral target site usage showed that chromatin assembly significantly influences integration by MLV and HIV-1 purified integrases or PICs isolated from MLV-infected cells. Integration into nucleosomal DNA was strongly preferred at certain sites, between 10- and 50-fold greater than the same sites in DNA devoid of proteins (Pryciak and

Varmus, 1992). The preferred sites on the nucleosomal DNA have a 10-base-pair periodic spacing, suggesting that these sites are located on the same face of the double helix. Comparison of the MLV integration sites with DNase I cleavage sites on a rotationally phased mononucleosome indicates that integration occurs preferentially at positions where the major groove is on the exposed face of the nucleosomal DNA (Pryciak and Varmus, 1992; Pryciak *et al.*, 1992b).

Since the nucleosomal complex usually restricts the accessibility of other proteins to the DNA, the enhancement of retroviral integration into nucleosomal DNA was unexpected. Chromatin assembly requires the wrapping of DNA around a histone octamer that results in the widening of both the major and minor grooves and an increased curvature of the exposed face of the histone-bound DNA (Richmond *et al.*, 1984). These changes in the local DNA structure create favorable sites for integration (Figure 2.3).

Are the histones in nucleosomal DNA attracting the integration machinery to those DNA sites, or is the histone-induced bending of DNA positively affecting integration? To distinguish the role of DNA bending and histones during integration, the N-terminal tails of the histones were removed without disrupting their globular domain, and integration by HIV-1 integrase was examined (Pruss *et al.*, 1994a). Removal of the histone tails is believed to allow the integration machinery better access to the histone. However, similar distributions and frequencies of integration were observed into intact or tailess histone-bound DNA. Therefore, the wrapping of DNA around nucleosomes is the determinant for integration site selection, and not the presence of a protein on the DNA. Enhancement of integration at different positions around the nucleosome was analyzed and shown to correlate positively with an increase in DNA distortion (Pruss *et al.*, 1994a).

To further examine the effect of DNA bending induced by histones and nonhistone proteins, the *E. coli* catabolite activator protein (CAP) and the Lac repressor (Muller and Varmus, 1994) were used in integration assays. In *in vitro* analyses, CAP- and Lac repressor-bound DNA indicate that DNA bending, regardless of whether a protein is bound to the curved sequence, creates favorable sites for integration. The CAP protein makes a ~90° bend at its DNA-binding site, and the Lac repressor induces DNA bends when bound to two of its operators by looping out an intervening spacer sequence. Lac operator constructs of two different lengths of the spacer, one where the loop was bent ~90° and the other ~30° per 25 base pairs, also served as targets for integration to help ascertain the effect of the degree of bending. Data from this experiment showed a preference of integration into both the CAP-induced bent DNA and the Lac repressor-induced looped DNA, with the degree of bending correlating with the stimulatory effect on integration.

Distortion of DNA by the formation of nucleosomes can also override the effect of intrinsic DNA structure on integration (Pruss *et al.*, 1994b). Regardless of whether the DNA is of a rigid and straight, flexible, or curved structure (see Section II.A.3), a similar alteration in the enhancement and distribution of integration is observed as with naturally assembled nucleosomes (Pruss et al., 1994b). The preferred sites have a periodicity of about 10 base pairs, which is in congruence with previous studies suggesting that the repetitive helical structure of DNA wraps around nucleosomes, and renders the same face of DNA available for integration (Pryciak et al., 1992b).

Although DNA bending creates attractive sites for retroviral cDNA insertion, accessibility to the DNA is also a key determining factor. Protein binding may bend DNA, but it can simultaneously block the availability of its cognate binding sites for integration. The effects of the DNA-binding domain of the human lymphoid enhancer factor (LEF) and the E. coli integration host factor (IHF) illustrate this point (Bor et al., 1995). Both proteins bend DNA to a similar extent and widen the major groove, but they have different effects on integration. Preferred integration is seen only with IHF-bound DNA, and the difference is attributed to the different conformations of the two protein–DNA complexes. IHF binds DNA on the inner face of the bent DNA, and the outer curved face is available for integration. LEF, on the other hand, binds on the outer face of the bent DNA (Love et al., 1995; Yang and Nash, 1994) and blocks accessibility of the integration machinery to the widened major groove.

2. DNA-binding proteins

In addition to bending DNA, DNA-binding proteins may attract the integration machinery to certain DNA regions through protein–protein interactions. The best-characterized example is the closely related yeast retrotransposons Ty3, which integrate into positions directly upstream of transcription start sites of genes transcribed by RNA polymerase III (Pol III) (Chalker and Sandmeyer, 1992; Kirchner et al., 1995). This integration preference is mediated by a specific interaction between the Ty3 integration machinery and Pol III transcription factors (Kinsey and Sandmeyer, 1991). In the mammalian system, Ini1 (integrase interactor 1), a cellular protein with sequence homology to the yeast transcription factor SNF5 of the SWI/SNF complex (Laurent and Carlson, 1992; Laurent et al., 1990), interacts specifically with HIV-1 integrase and enhances integration in in vitro assays (Kalpana et al., 1994). A specific interaction between retroviral integrase and a transcriptional regulatory protein may provide a mechanism for targeting integration to open chromatin regions, consistent with earlier reports that MLV integrates preferentially into regions near DNase I-hypersensitive sites and transcriptionally active genes (Rohdewohld et al., 1987; Vijaya et al., 1986). Presently, it is not known whether this type of protein–protein interaction is a common mechanism used by retroviruses for the selection of integration sites.

In some cases, the presence of DNA-binding proteins can block integration to their recognition sites by steric hindrance. Studies with several transcriptional repressors, such as the yeast Matα2, *E. coli* Lac repressor (Muller and Varmus, 1994) and LexA repressor (Goulaouic and Chow, 1996), and the phage λ repressor (Bushman, 1994), showed that integration is blocked at their cognate DNA-binding sites. Similarly, the centromeric alphoid repeats, a highly compact and likely inaccessible region of the chromosomal DNA, are devoid of integration events (Carteau *et al.*, 1998).

3. DNA sequence and structure

Several studies indicate that the sequence of DNA may be a factor for the choice of integration sites by retroviruses (Carteau *et al.*, 1998; Pryciak and Varmus, 1992; Stevens and Griffith, 1996). Evidence linking the importance of DNA sequence to integration preference is currently weak, as a *bona fide* consensus DNA sequence for retroviral integration has not been uncovered (Bor *et al.*, 1996; Fitzgerald and Grandgenett, 1994; Goodarzi *et al.*, 1997). However, sequence determination of viral-target DNA junctions suggests a bias for certain bases. For HIV-1, a 5-base-pair preferred sequence, 5'-GT(A/T)(A/T)(T/C)-3', flanking the provirus has been identified (Carteau *et al.*, 1998; Stevens and Griffith, 1996). For ALV, integration is favored 5' of G or C residues (Fitzgerald and Grandgenett, 1994).

The ability of virtually any DNA sequence to serve as an integration site and the identification of only weak consensus sequences among the target sites indicate that the base identity per se is not a major determinant. Instead, it suggests that local DNA structures determined indirectly by the base sequence influence the usage of a particular site (Muller and Varmus, 1994; Pruss *et al.*, 1994b). Integration into a 50-base-pair DNA segment of different intrinsic structures was studied using purified HIV-1 and MLV integrases (Pruss *et al.*, 1994b). Rigid and straight DNA was formed using oligo(dA)·oligo(dT) tracts and flexible DNA was formed using oligo[d(A-T)] tracts. A curved DNA segment, which mimics the naturally curved kinetoplast DNA, contained 5 base pairs of oligo(dA)·oligo(dT) tracts every 10–11 base pairs separated by GGCC/CCGG base pairs (Brukner *et al.*, 1993). DNA with intrinsic curvature serves as the best target substrate for integration over rigid and flexible DNA, with rigid DNA being the least favorable target (Pruss *et al.*, 1994b). The curved DNA segment has a 10–11 base pairs periodic enhancement of integration into the GGCC sequence and the immediately adjacent TpA step. This is a highly curved region (Goodsell *et al.*, 1993) with its widened minor groove lying along the outer face of a curved DNA (Satchwell *et al.*, 1986), thereby providing further evidence to the important effect of DNA bending on integration.

Enhanced integration into DNA with a specific secondary structure is seen with purified ASV and HIV-1 integrases using a supercoiled DNA containing an extensive inverted repeat (Katz *et al.*, 1998). The preferred integration is on only one half of the inverted repeat and thus suggests that primary sequence is not the basis for the integration bias. Rather, the inverted repeat forms a secondary structure resembling a cruciform or hairpin loop. The enhanced integration maps to a small region directly adjacent to the loop, a preference not observed with the linear version of this target. The reason for this bias may be due to the resemblance of the end loop to partially unwound DNA, and perhaps is a more accessible and energetically favorable target for integration (Katz *et al.*, 1998).

Previous findings on target DNA with intrinsic or secondary structure appear to reflect a necessity of integrase to distort DNA or unstack base pairs as part of its reaction mechanism during integration. In 3′-end processing, disruption of base pairing at the viral DNA terminus is a required step during catalysis (Scottoline *et al.*, 1997). Since both 3′-end processing and 3′-end joining reactions are apparently carried out by a single active site, perhaps integrase also prefers partial unwinding of target DNA during strand transfer.

4. Other factors affecting integration

Previous studies on retroviral integration suggest that integration is favored within transcriptionally active regions of the genome (Rohdewohld *et al.*, 1987; Scherdin *et al.*, 1990; Vijaya *et al.*, 1986). A small survey of MLV integration recombinants, using Southern blot analysis and nuclear run-on experiments, determined that viral gene insertion is statistically biased near DNase-I hypersensitive sites and some of these sites are transcriptionally active (Rohdewohld *et al.*, 1987; Scherdin *et al.*, 1990; Vijaya *et al.*, 1986). This phenomenon seems plausible since this DNA region is often in an unwound or "open" conformation and, therefore, may provide a more desirable site for integration and ensure subsequent transcription of the proviral DNA. It has also been proposed that transcriptionally active regions may be preferred for integration due to the distortion of B-DNA into a structure similar to Z-DNA (Wittig *et al.*, 1991, 1992), which has a considerably wide major groove and may provide a more desirable site for integration.

It has been difficult to ascertain the role of transcription during integration because transcriptionally inactive regions have also been shown to be available for integration (Carteau *et al.*, 1998; Kitamura *et al.*, 1992). Incubation of ALV-infected cell extracts with a target plasmid containing methylated cytosines in CpG islands (Kitamura *et al.*, 1992) revealed a strong bias for integration into these methylated sequences. Methylation of cytosines is a regulatory mechanism correlated to transcriptional inactivity or "silent" regions of chromosomes. The observed integration preference into methylated CpG regions may be due to methylation-induced changes in local structure from B- to Z-DNA (Behe et al., 1981; Kitamura et al., 1992).

One possible explanation for preferred integration into both methylated CpG islands and transcriptionally active regions may be due to their similar distortions of the DNA structure to a form that resembles Z-DNA. In vivo data of HIV-1 integration within the human genome showed that transcriptionally active regions are not necessarily preferred for integration and that most of the chromosome is available for integration except for the compact, centromeric alphoid regions (Carteau et al., 1998).

In addition to transcription and DNA methylation, experiments conducted with HIV-1-infected cells showed that integration may be favored within 600–800 base pairs of repetitive elements, Alu or LINE-1 (Stevens and Griffith, 1994, 1996), or topoisomerase II cleavage sites (Howard and Griffith, 1993). The basis for such preferences is not clear, and the result is inconclusive. In the case of repetitive sequences, another study did not find strong biases either for or against integration near Alu or LINE-1 elements (Carteau et al., 1998).

It is evident that further in vivo analyses of a larger pool of integration recombinants are needed to decipher the influence of DNA features, chromosomal structures, and protein factors on target site selection. The analyses of large numbers of integration sites by current methods are time-consuming and labor-intensive and a high-throughput method will prove valuable for studying retroviral integration in vivo.

B. Role of integrase in target site selection

In addition to DNA-binding proteins and local structural features of DNA, integrase itself plays a key role in target site selection. Although integrase can use many sites along a chromosome for integration, HIV-1, FIV, MLV, and visna virus integrases exhibit significant differences in the distribution and frequency of integration into an identical target substrate. Since different integrases generate unique and reproducible integration patterns (Figure 2.4; Bushman and Wang, 1994; Katzman and Sudol, 1995; Pryciak et al., 1992a; Shibagaki and Chow, 1997), a motif determining target site selection must exist within integrase. Currently, the mechanism of target site recognition and selection by integrase is poorly understood. It is of particular interest to identify the regions of integrase involved in protein–DNA interactions to provide a further understanding of target site selection.

An informative approach to mapping the target site determinant within integrase is to analyze the integration site preference of chimeric proteins (Katzman and Sudol, 1995; Shibagaki and Chow, 1997). Since different integrases produce different integration patterns, as defined by the distribution and

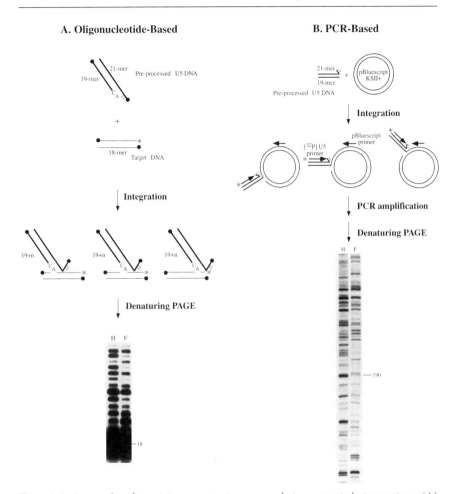

Figure 2.4. Assays for determining target site usage during retroviral integration. (A) Oligonucleotide-based assay. The donor U5 DNA (thick lines) is unlabeled and the target DNA (thin lines) is labeled at the 3′-end of one DNA strand (asterisk). The 3′-end joining reaction generates labeled products that are longer $(19 + n)$ than the input substrate. The product lengths are heterogeneous because the cleavage-joining sites are nonspecific. (B) PCR-based assay. The joining reaction includes the preprocessed viral DNA and the supercoiled Bluescript plasmid DNA. Thick arrows denote the primers for the PCR reaction, and the asterisk indicates the 5′-end labeled primer. (+) and (−) denote the plus and minus strands, respectively, of the plasmid DNA. In both (A) and (B), the reaction products were separated on a denaturing polyacrylamide gel. Numbers on the right of the panel indicate lengths in nucleotides of size markers. H and F denote that the reactions were carried out in the presence of HIV-1 and FIV integrases, respectively.

frequency of integration events, swapping the target site-specifying domain should result in a corresponding exchange of characteristic integration patterns between the resulting chimeras. This method depends on exchange of integration patterns for identification of the target site-specifying domain. Therefore, the chimeric approach avoids possible complications, such as conformational changes and other secondary effects, introduced by mutagenesis and deletion analyses.

One study generated HIV-1/FIV chimeric proteins by swapping each of the three domains of integrase, the N-, the C-terminus, and the core. The target site-specifying domain was then mapped by comparing the patterns of integration by the chimeric proteins using oligonucleotide- and PCR-based integration assays (Shibagaki and Chow, 1997). The oligonucleotide-based assay (Figure 2.4A) uses a short duplex DNA resembling the HIV U5 LTR as a substrate for integration into a 3'-end-labeled target DNA (Leavitt *et al.*, 1992; Shibagaki and Chow, 1997), and the more sensitive PCR-based integration assay (Figure 2.4B) involves the amplification of integration events along a target plasmid (Chow, 1997; Pryciak and Varmus, 1992). In the PCR-based assay, integration is first allowed to proceed by incubating integrase, short oligonucleotide substrates that mimic the viral LTR end, and a target plasmid DNA. PCR is then performed on the integration products with a radiolabeled 5' primer that anneals to the viral LTR and a 3' primer that anneals to a specific location on the target plasmid DNA. The radioactive PCR products are separated on a denaturing gel, and the precise sites where integration occurred can be determined by mapping the size of the product bands against the location of the 3' primer for the plasmid. Moreover, the intensity of the bands on the gel is proportional to the frequency of integration into a particular target site. An overview of the ladder of integration events represents an integration pattern that is characteristic of each integrase.

Integration data from the HIV-1/FIV chimeric proteins revealed that an exchange of either the N- or C-terminus does not influence target site selection. Rather, all chimeras with an HIV-1 core domain (amino acid residues 50–234) exhibit an HIV-1 integration pattern and all chimeras with an FIV core domain (amino acid residues 52–235) exhibit an FIV integration pattern. These results indicate that the target site determinant resides within the core domain of integrase.

Although target site selection is mapped to the core domain, further narrowing of the target site determinant is necessary to define the minimal motif and provide a better understanding of the protein–DNA interaction. The identified target site-specifying domain includes a portion of the C-terminus involved in nonspecific DNA binding (amino acid residues 213–266) (Engelman *et al.*, 1994), and the previously described HIV-1/FIV chimeras were not able to discern if this DNA-binding domain contributed to target site selection. New HIV-1/FIV chimeric integrases were made and analyses of the integration patterns mapped the target site determinant within residues 50–186 of the core domain (Appa and

Chow, unpublished results). This finding indicates that the nonspecific DNA-binding region is not involved in target site selection.

A similar conclusion was reached from nonspecific alcoholysis data using chimeric integrases generated from HIV-1 and visna virus. Nonspecific alcoholysis is an endonuclease activity of integrase that uses glycerol or water instead of the 3′-OH end of the viral DNA as the nucleophile (Katzman and Sudol, 1995). Since integrases of HIV-1 and visna virus produced different cleavage patterns of nonspecific alcoholysis (Katzman and Sudol, 1995), chimeras of these two proteins were used to map the target site-specifying domain. Residues 50–190 are identified to be responsible for target DNA recognition (Katzman and Sudol, 1995). However, it is questionable whether the motif conferring target DNA recognition in nonspecific alcoholysis is identical to that responsible for target site selection during integration.

As discussed earlier, the core domain of HIV-1 can bind the disintegration Y-mer substrate (Engelman et al., 1994) and certain peptides within the core can crosslink the target DNA portion of the dumbbell disintegration substrate (Heuer and Brown, 1997). It is presently not known if the region capable of binding the Y-mer substrate is identical to or overlaps the target site selection domain. Due to a lack of structural data on full-length integrase and detailed biochemical data on protein–DNA interactions, a model of full-length integrase with viral and target DNA is not available. However, a model consisting of integrase's core and carboxy domains complexed with viral and target DNA has been generated by compiling and fitting known biochemical and structural data on integrase (Heuer and Brown, 1998). This model predicts that a portion of the $\alpha 3$ and $\alpha 4$ helices within the core domain may be in close proximity to target DNA. In particular, amino acids S153 and K160 within the $\alpha 4$ helix have been suggested to interact with target DNA (Heuer and Brown, 1998). Based on this model, it seems reasonable to hypothesize that one or both of the $\alpha 3$ and $\alpha 4$ helices are involved in target site selection, as they both reside within residues 50–186. Further studies using the described chimeric approach or other biochemical and structural analyses may resolve this issue.

Thus far, target site selection has been studied in vitro using purified integrases and PICs, but it is not known if integrase plays a similar role in vivo. Comparing the integration patterns of purified HIV-1 integrase and PICs isolated from HIV-1-infected cells revealed a difference in target site usage (Bor et al., 1996). However, in the case of either MLV or ALV, target selection was similar, regardless of whether PICs or purified integrase were the source of integration activity (Kitamura et al., 1992; Pryciak and Varmus, 1992). Questions as to whether other viral proteins in the PIC contribute to target site selection or if different integrases have different requirements for in vivo target site selection are still unanswered.

Studying target site selection will enable us to further understand protein–DNA interactions, and this knowledge may be used to improve the specificity of retroviruses as a vector for genetic engineering or gene therapy. One drawback of using retroviral vectors is the occurrence of insertional mutagenesis due to the nonspecific nature of integration. Information on how integrase or the integration complex recognizes and selects target sites may help to develop retroviral vectors with target site specificity. Approaches to achieving site-directed integration will be discussed next.

III. SITE-DIRECTED INTEGRATION

The ability of retroviruses to permanently and precisely insert their genome into the chromosome of an infected cell is a desirable property that can be exploited for the purpose of gene therapy. However, the nonspecific nature of integration can be a potential pitfall for introducing a gene of interest with retroviral vectors (Carteau et al., 1998; Hughes et al., 1978; Withers-Ward et al., 1994). By insertional mutagenesis, integration could inactivate an essential host cellular gene, or inappropriately cause overexpression of an undesirable gene, such as a proto-oncogene, because of regulatory elements that reside in the ends of the viral genome. To alter the site specificity of retroviral integration, a fusion protein strategy has been established. The fusion protein consists of a sequence-specific DNA-binding protein and a retroviral integrase (Bushman, 1994; Bushman and Miller, 1997; Goulaouic and Chow, 1996; Katz et al., 1996). The sequence-specific DNA-binding protein directs integration by recognizing and binding to its target site on the DNA, causing integration to be mediated into the immediately adjacent region, which is available to integrase.

A. Directed integration using integrase-fusion proteins *in vitro*

To initially determine if integrase-fusion proteins could be used to direct integration into specific DNA sequences, proteins were expressed and purified from *E. coli*, and tested in *in vitro* integration reactions. The proteins examined include a fusion of the DNA-binding domain (DBD) of phage λ repressor to the N-terminus of HIV-1 integrase (Bushman, 1994), a fusion of the full-length *E. coli* LexA repressor or its DBD to the C-terminus of HIV-1 integrase (Goulaouic and Chow, 1996), and a fusion of the LexA DBD to the N- or C-terminus of ASV integrase (Katz et al., 1996). Both the λ and LexA repressors are small proteins, around 20 kDa, that consist of an N-terminal DNA-binding domain and a C-terminal dimerization domain. Each binds to a 16–17 base-pair consensus sequence with dyad symmetry and, depending on the sequence of the binding site, can have a K_m as low as 2×10^{-10} M. Fusion of the repressor proteins to the N- or C-terminus of HIV-1 or ASV integrase does not affect the cognate integration activity of the purified integrase-fusion protein relative to the wild-type protein (Bushman, 1994; Goulaouic and Chow, 1996; Katz et al., 1996).

Further examination of these fusion proteins revealed an ability to direct integration toward the binding site, or operator, of the repressor. This determination was made using an agarose gel-type assay (Bushman, 1994; Goulaouic and Chow, 1996), and a PCR-based assay (Chow, 1997; Pryciak and Varmus, 1992). The agarose gel assay tests the frequency of integration by the fusion proteins into large fragments of DNA, which may or may not contain the target sequence for the repressor. It is performed by first digesting a plasmid containing the binding site with a restriction enzyme, generating 6–10 DNA fragments. The integrase-fusion protein is then incubated with the digested plasmid and radiolabeled oligonucleotides that mimic the U5 end of the viral genome. Because the viral DNA is radiolabeled, the fragments into which integration took place can be detected by autoradiography. When HIV-1 integrase was fused to either the DBD of λ repressor or the full-length LexA repressor, the fragment containing the operator site for the repressor protein was preferentially selected for integration (Bushman, 1994; Goulaouic and Chow, 1996). A 14-fold greater integration frequency was observed by the HIV-1 integrase-LexA fusion protein into the DNA fragment containing the LexA operator when the fusion protein and digested plasmid were present in the reaction at a ratio of 1:1 (Goulaouic and Chow, 1996). The HIV-1 integrase-λ repressor DBD protein selected the fragment containing the λ repressor-binding site at least as well (Bushman, 1994). In contrast, when wild-type HIV-1 integrase was used to catalyze integration, each fragment was detected with the same intensity on the gel after autoradiography, indicating equal usage by integrase. Likewise, when either fusion protein was incubated with a digested target plasmid that lacked the repressor-binding site, each fragment was used equally as target DNA, demonstrating that the bias previously observed depends on the repressor-binding site on the target DNA (Bushman, 1994; Goulaouic and Chow, 1996).

A similar agarose gel assay was used to determine whether ASV integrase, fused at the N-terminus to the LexA DBD, was capable of site-directed integration. A plasmid DNA that contained the LexA operator was linearized and then incubated with the integrase-LexA DBD fusion protein. Because ASV integrase has relatively strong endonuclease activity, it can introduce nicks that result in double-standed breaks in DNA. The site of fusion protein binding can then be deduced from the size of the cleaved DNA products. When the fusion protein was incubated with the linearized plasmid containing the LexA-binding site, DNA fragments migrated at the size expected for nicking at the LexA operator. However, when the wild-type ASV integrase was used to catalyze the reaction, a distinct and detectable nicking pattern was no longer seen (Katz et al., 1996).

While both of these assays demonstrate that integration occurs within the greater vicinity of the sequence for the site-specific DNA-binding protein, they do not reveal the exact phosphodiester bond into the target DNA where integration takes place. The PCR-based assay, described earlier (see Section II.B and Figure 2.4B) for determining the distribution and frequency of integration events, was performed to address this issue. When HIV-1 integrase was fused to

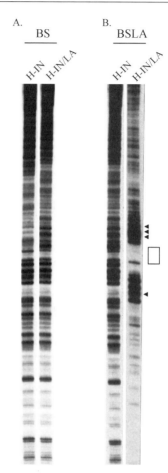

Figure 2.5. Site-directed integration mediated by HIV-1 integrase and HIV-1 integrase-LexA fusion proteins. (A) Integration patterns of purified proteins into a control plasmid. (B) Site-directed integration by HIV-1 integrase-LexA into the plasmid containing the LexA operator (open box) using the PCR-based assay (see Figure 2.4B). Integration enhancement sites flanking the operator are denoted by filled arrows. BS, pBluescript; BSLA, pBluescript with the LexA operator sequence; H-IN, HIV-1 integrase; H-IN/LA, HIV-1 integrase fused to LexA.

either the LexA repressor or the DBD of phage λ repressor, or ASV integrase is fused to the LexA DBD, a shift in the pattern of integration from wild-type integrase was detected in the presence of the repressor-binding site (Figure 2.5). While the wild-type integrases produced a nonspecific pattern of integration throughout the plasmid, the integrase-fusion proteins generated a pattern that exhibited several distinct characteristics. First, the location on the gel corresponding to the site of the operator for LexA lacked bands, indicating that it is no longer used for integration. Second, the DNA sites that immediately flank the operator sequence for the repressor protein were dramatically increased in number and intensity. Finally, the intensity of the bands not in the vicinity of the operator sequence were diminished compared to wild-type integrase. These results likely indicate that the repressor component of the fusion protein recognizes and binds to its specific sequence on the target DNA, occluding it from integration. Because the local concentration of the fusion protein is increased at the binding site, DNA regions near the operator are selectively used for integration. In addition, sequestration of the fusion protein at the repressor-binding sequence results in fewer integrases that are available to mediate integration into the DNA outside the proximity of the operator. Consistent with this model is the observation that as the concentration of the fusion protein relative to the target plasmid was increased, a loss of site-specific integration was observed (Goulaouic and Chow, 1996). Also, when the fusion proteins were used with a target that did not contain the operator sequence for the repressor, site specificity was lost and a nonspecific integration pattern was produced (Bushman, 1994; Goulaouic and Chow, 1996; Katz *et al.*, 1996).

Although it is apparent that fusion of a sequence-specific DNA-binding protein to a retroviral integrase can direct integration into specific sites, it is also noteworthy to point out that there remains integration into the target DNA outlying the operator sequence. Therefore, the importance of identifying the integrase domain responsible for binding nonspecific DNA and for target site selection should be emphasized (see Section II.B).

B. Incorporating integrase-fusion proteins into infectious virions

A direct and simple way to introduce a gene of interest into a desired mammalian cell is by transduction with retrovirus-based vectors. To manipulate integration sites by these vectors, integrase-fusion proteins need to be incorporated into retroviruses, the viruses must be able to infect cells, and the fusion proteins in them must be functional to mediate integration into the desired sites. Thus far, attempts to include an HIV-1 integrase-zif 268 DBD (Bushman and Miller, 1997) and ASV integrase-LexA DBD (Katz *et al.*, 1996) fusion protein into their respective viruses have been made. In each case, the gene encoding the sequence-specific DNA-binding protein was fused at the 3′-end of the integrase gene in the viral genome. Neither attempt was able to achieve expression of the integrase-fusion protein and retain the ability of the resulting retrovirus to infect cells. The HIV-1 containing integrase-zif 268 DBD expressed the fusion protein to detectable levels by Western blot analysis, but was unable to infect an indicator cell line. To generate infectious viruses with the integrase-zif 268 viral clone, it was necessary to cotransfect producer cells with an expression construct for the wild-type virus (Bushman and Miller, 1997). The ASV virions containing the integrase-LexA DBD presented a different problem. After transfection of the recombinant viral clone, low titers of virus were produced. Thirty days posttransfection, when kinetics similar to the wild-type virus were observed, integrase-specific immunoblots detected the integrase-LexA DBD fusion protein at a level of less than 20% of total immunoreactive products in the transfected cell cultures. No fusion protein was found in the virus particle harvested from the culture medium. It is likely that inclusion of the integrase-LexA DBD fusion gene negatively affected viral replication, and subsequent emergence of replicating virus might represent a reversion event (Katz et al., 1996).

The exact reason for the loss in fusion protein production and infectivity is not known. A potential cause is the complicated and delicate coding region at the 3′-end of the integrase gene in the viral genome. For HIV-1, the 3′-end of the integrase gene overlaps the viral accessory gene vif, and there is an important splice acceptor site embedded in this region (Figure 2.2; Purcell and Martin, 1993). Introduction of the zif 268-coding sequence interrupted not only the vif gene but also that of another accessory protein, vpr (Bushman and Miller, 1997). Although part of this disruption was rescued by reinserting these two genes back into the viral clone, interference with the location of a viral splice acceptor site remained. The 3′-end of the integrase coding sequence in the ASV clone also presents several potential pitfalls. The 3′-end of the integrase gene in ASV overlaps the envelope reading frame by a 37-amino acid coding sequence. A splice acceptor site for the envelope transcript is also present after the integrase coding sequence, and a protease cleavage site follows the integrase gene. Although the interrupted sequences were also reintroduced, the attempt was unsuccessful in leaving the virus unperturbed (Katz et al., 1996).

Despite the difficulties observed when the integrase-zif 268 DBD gene was inserted into the viral genome, the HIV-1 containing an integrase-zif 268 DBD protein was able to produce infectious virus particles after cotransfection with a wild-type viral genome. This allowed investigation into the ability of the zif 268 DBD to influence integration site selection using PICs isolated from cells infected with this virus. PICs, described earlier in this chapter, are obtained by infecting susceptible cells at a high viral titer and then lysing them at the appropriate time, usually 6–8 h after infection, when formation is complete but before entry into the nucleus. Incubation of the PICs with a target plasmid DNA in vitro results in authentic two-ended integration of the viral cDNA into the target plasmid. The PCR-based assay can then be carried out to determine the sites of integration. In the presence of a target plasmid that contained the zif 268-binding site, PICs isolated from cells infected with the fusion protein-containing virus displayed several features described earlier for the purified integrase-fusion proteins in the same assay. Integration into the zif 268-binding site was decreased, while

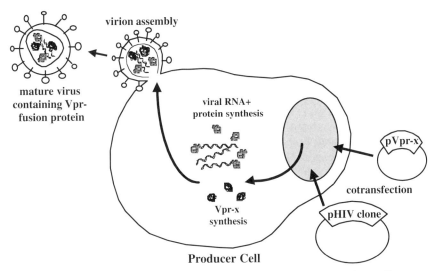

Figure 2.6. Schematic representation of the *in trans* incorporation method. Producer cells are co-transfected with a viral expression construct (pHIV) and the Vpr-fusion protein vector (pVpr-x). After protein expression, the Vpr-fusion protein (Vpr-x) is delivered into the virus particles via the interaction of Vpr with the p6 portion of Gag.

DNA sites in close proximity to the zif 268-recognition sequence was increased (Bushman and Miller, 1997). Although the influence over the integration pattern by these PICs was not as pronounced as that of the purified proteins in the PCR-based assay, it is important to keep in mind that the PICs contained a mixture of wild-type integrase and the integrase-zif 268 DBD protein. Therefore, it is as yet difficult to determine the efficacy of a fusion protein to mediate site-directed integration in a retroviral vector.

Recently, a new method for incorporating proteins into HIV-1 has been established that is useful for including integrase-fusion proteins into viral particles. This method is termed *in trans* incorporation because the protein to be packaged in the virus is not encoded by the viral genome, but rather is expressed off a separate DNA construct by the packaging cell line after transfection (Figure 2.6). Several features of this *in trans* expression construct include an HIV-2 LTR at the 5'-end of the gene of interest to drive its expression, an HIV-2 RRE located 3' to the gene to ensure that the viral transcripts are exported by Rev to the cytoplasm for expression, and a polyadenylation signal stabilizing the transcript and increasing translational efficiency (Wu *et al.*, 1995). The protein encoded between the HIV-2 elements is a Vpr-fusion protein. Vpr is at the N-terminus of the fusion protein and acts as a shuttle to deliver the desired protein into HIV-1. Vpr has many functions during retroviral infection, but the property that is exploited by the *in trans* incorporation method is its ability to interact with the p6 portion of Gag during virus assembly and budding (Kondo and Gottlinger, 1996; Liu *et al.*, 1997a). Fusing an exogenous protein to the C-terminus of Vpr will cause its incorporation into HIV-1 particles via this interaction. A number of Vpr-fusion proteins have been successfully incorporated into HIV-1 (Fletcher *et al.*, 1997; Liu *et al.*, 1997a, 1999; Wu *et al.*, 1995, 1997, 1999), including a Vpr-integrase fusion protein (Fletcher *et al.*, 1997; Liu *et al.*, 1997a; Wu *et al.*, 1997). Viruses containing a mutant integrase alone are noninfectious, but when provided with the Vpr-integrase protein *in trans* gain the ability to infect susceptible cells at about 20% of the wild-type virus. For the Vpr-integrase fusion protein to complement integration activity to the integrase mutant virus, Vpr must be removed from integrase by HIV-1 protease cleavage after incorporation (Fletcher *et al.*, 1997).

Similarly, an integrase fused with a sequence-specific DNA-binding protein can be incorporated into HIV-1 by the *in trans* method, with the advantage of avoiding interference with critical elements at the 3'-end of the integrase coding region of the viral genome. This strategy has been used to provide an integrase-LexA protein *in trans* to an HIV-1 genome containing a D64V mutation in integrase. D64 is one of the catalytic core triad residues essential for catalysis. After cotransfection of producer cells with a construct expressing the integrase-mutated genome and a construct encoding the fusion protein, immunoblotting has indicated the presence of the Vpr-integrase-LexA protein in HIV-1 particles and the proteolytic cleavage of Vpr from integrase-LexA during virus maturation. Testing for viral infectivity and expression has revealed that the integrase-mutated viruses containing the integrase-LexA fusion protein are able to express reporter genes after infecting target cells, whereas integrase-mutated viruses alone cannot. The *in trans* method has already enabled the production of virus particles that contain a high level of the integrase-LexA fusion protein and that are infectious, a previous obstacle of encoding the integrase-LexA fusion protein in the viral genome (Holmes-Son and Chow, unpublished results). The remaining question to be addressed is the efficiency of the integrase-LexA fusion protein to direct integration into its operator sequence.

C. Future direction: site-specific retroviral vectors

The development of retroviral vectors that contain an integrase-fusion protein to direct integration of transgenes into predetermined sites is promising. The usefulness of this strategy depends on the feasibility of eliminating nonspecific DNA binding by the integrase component and increasing the site specificity by the DNA-binding component of the fusion protein. Defining the role of integrase in choosing integration sites is currently being worked out by several groups (Katzman and Sudol, 1995, 1998; Shibagaki and Chow, 1997), and has been covered in previous sections of this chapter. It is currently not known whether the integrase domain responsible for target site selection, if it is identified, can be manipulated for targeting purposes. It is also unknown if removing the ability of integrase to bind nonspecific DNA will also negatively affect catalysis. Therefore, a more immediate prospect toward preparing a site-directed retroviral vector is to test different fusion partners with integrase for their ability to confer high specificity toward particular sites on the chromosomal DNA.

One attractive class of DNA-binding proteins is the family of Cys_2-His_2 zinc-finger proteins, of which zif 268 is a member. These proteins offer several advantages for use in conferring site specificity to retroviral integrases. First, they can be designed to recognize novel DNA sequences. The residues of the protein that make contact with the DNA are known, from structural analysis of the zif 268-DNA complex (Pavletich and Pabo, 1993). Mutagenesis of strategic amino acids followed by phage display selection can yield novel zinc-finger proteins with binding sites for desired DNA sequences (Rebar and Pabo, 1994). In addition, fusion of two of these zif proteins together yields a protein with an 18-base-pair-binding site, sufficient to recognize any unique site in the human genome (Beerli et al., 1998; Kim and Pabo, 1998; Liu et al., 1997b; Wang and Pabo, 1999). Furthermore, because these proteins are put through multiple rounds of selection for their target sequence, they are highly specific, with dissociation constants in the femtomolar range being reported (Kim and Pabo, 1998).

A second type of integrase-fusion protein strategy could take its lead from the Ty family of retrotransposons found in *Saccharomyces cerevisiae*. The Ty family of retrotransposons has five family members, and each mediates integration with high specificity into the yeast genome. Ty elements 1–4 integrate nearly exclusively within 750 base pairs of the 5′-end of genes transcribed by Pol III (Chalker and Sandmeyer, 1990, 1992; Devine and Boeke, 1996; Ji et al., 1993; Kim et al., 1998), and Ty5 insertions are always located in the telomeric chromatin, or at the mating loci (Kim et al., 1998; Zou et al., 1996; Zou and Voytas, 1997). The mechanism of this site-directed integration, at least in the case of Ty3 and Ty5, is through protein–protein interactions. As discussed earlier, Ty3 interacts directly with the Pol III transcription factor B/C complex, causing precise integration within 2 bases of the start of transcription (Chalker and Sandmeyer, 1992), and Ty5 recognizes a component of the telomeric chromatin (Zou and Voytas, 1997). The high specificity of integration resulting from protein–protein interactions may also be adapted in the construction of targeted retroviral vectors. Rather than fuse integrase to a sequence-specific DNA-binding protein, integrase may be fused to a protein that recognizes a DNA-binding factor, such as a component of the human Pol III transcription complex. One example is the retinoblastoma protein, which interacts directly with TFIIIB and TFIIIC2 (Larminie et al., 1997). Benefits of this strategy are that most of the Pol III-transcribed genes are present in multiple copies. Therefore, the integrase-fusion protein will have improved ability to locate its target, and a decreased risk of disrupting normal cellular function. Integration of retroviral vector DNA into Pol III-transcribed genes may also direct the transgene into a transcriptionally active region of the genome, enhancing the chance of transgene expression. Finally, as hoped in the case of Ty elements, tethering integrase through protein–protein interactions may provide higher specificity than that of protein–DNA interactions.

IV. CONCLUDING REMARKS

The continued study of retroviral integrases will contribute to the understanding of retrovirology and advance the development of gene therapy vectors. It is evident from this chapter that there are many aspects of integration that remain poorly understood, ranging from a full structure of the enzyme to its active multimeric state and the mechanism of 5′-end joining, a critical step in the viral integration process. Resolving these questions will give us better information to comprehending how the genetic information of retroviruses is inserted into target cells and how retroviruses replicate. Also, integration is a critical component toward the use of retroviruses in gene therapy. The ability of retroviruses to precisely and permanently introduce foreign genes into cellular chromosomes holds great promise for genetic engineering in higher eukaryotes and for gene therapy. The nonspecific nature by which integrase chooses integration sites may present a major drawback to the safety of these vectors. Efforts to determine the domain, or residues of integrase responsible for this function, and the means of conferring specificity to integration may improve the therapeutic application of current retroviral vectors.

Acknowledgments

This work was supported by National Institutes of Health Grant CA68859 to S.A.C. M.L.H.-S. was supported by a predoctoral fellowship from the Universitywide AIDS Research Program, University of California. R.S.A. was supported by a predoctoral fellowship from the UCLA AIDS Institute.

References

Asante-Appiah, E., Seeholzer, S. H., and Skalka, A. M. (1998). Structural determinants of metal-induced conformational changes in HIV-1 integrase. *J. Biol. Chem.* **273**, 35078–35087.
Asante-Appiah, E., and Skalka, A. M. (1999). HIV-1 integrase: Structural organization, conformational changes, and catalysis. *Adv. Virus Res.* **52**, 351–369.
Asante-Appiah, E., and Skalka, A. M. (1997). A metal-induced conformational change and activation of HIV-1 integrase. *J. Biol. Chem.* **272**, 16196–16205.
Balakrishnan, M., and Jonsson, C. B. (1997). Functional identification of nucleotides conferring specificity to retroviral integrase reactions. *J. Virol.* **71**, 1025–1035.

Beerli, R. R., Segal, D. J., Dreier, B., and Barbas, C. F., III. (1998). Toward controlling gene expression at will: Specific regulation of the *erbB2/HER-2* promotor by using polydactyl zinc finger proteins constructed from modular building blocks. *Proc. Natl. Acad. Sci. (USA)* **95**, 14628–14633.

Behe, M., Zimmerman, S., and Felsenfeld, G. (1981). Changes in the helical repeat of poly(dG-m5dC)·poly(dG-m5dC) and poly(dG-dC)·poly(dG-dC) associated with the B-Z transition. *Nature* **293**, 233–235.

Bor, Y.-C., Bushman, F. D., and Orgel, L. (1995). In vitro integration of human immunodeficiency virus type 1 cDNA into targets containing protein-induced bends. *Proc. Natl. Acad. Sci. (USA)* **92**, 10334–10338.

Bor, Y. C., Miller, M. D., Bushman, F. D., and Orgel, L. E. (1996). Target-sequence preferences of HIV-1 integration complexes in vitro. *Virology* **222**, 283–288.

Bowerman, B., Brown, P. O., Bishop, J. M., and Varmus, H. E. (1989). A nucleoprotein complex mediates the integration of retroviral DNA. *Genes Dev.* **3**, 469–478.

Brown, H., Chen, H., and Engelman, A. (1999). Structure-based mutagenesis of the human immunodeficiency virus type 1 DNA attachment site: Effects on integration and cDNA synthesis. *J. Virol.* **73**, 9011–9020.

Brown, P. O. (1997). Integration. *In* "Retroviruses" (J. M. Coffin, S. H. Hughes and H. E. Varmus, eds.), pp. 161–203. Cold Spring Harbor Laboratory Press, Cold Spring Harbor, NY.

Brown, P. O., Bowerman, B., Varmus, H. E., and Bishop, J. M. (1989). Retroviral integration: Structure of the initial covalent product and its precursor, and a role for the viral IN protein. *Proc. Natl. Acad. Sci. (USA)* **86**, 2525–2529.

Brukner, I., Dlakic, M., Savic, A., Susic, S., Pongor, S., and Suck, D. (1993). Evidence for opposite groove-directed curvature of GGGCCC and AAAAA sequence elements [published erratum appears in *Nucleic Acids Res.* (1993) **21**, 1332]. *Nucleic Acids Res.* **21**, 1025–1029.

Bujacz, G., Alexandratos, J., Qing, Z. L., Clement-Mella, C., and Wlodawer, A. (1996a). The catalytic domain of human immunodeficiency virus integrase: Ordered active site in the F185H mutant. *FEBS Lett.* **398**, 175–178.

Bujacz, G., Jaskolski, M., Alexandratos, J., Wlodawer, A., Merkel, G., Katz, R. A., and Skalka, A. M. (1996b). The catalytic domain of avian sarcoma virus integrase: Conformation of the active-site residues in the presence of divalent cations. *Structure* **4**, 89–96.

Bujacz, G., Jaskolski, M., Alexandratos, J., Wlodawer, A., Merkel, G., Katz, R. A., and Skalka, A. M. (1995). High resolution structure of the catalytic domain of the avian sarcoma virus integrase. *J. Mol. Biol.* **253**, 333–346.

Bukrinsky, M. I., Sharova, N., McDonald, T. L., Pushkarskaya, T., Tarpley, W. G., and Stevenson, M. (1993). Association of integrase, matrix, and reverse transcriptase antigens of human immunodeficiency virus type 1 with viral nucleic acids following acute infection. *Proc. Natl. Acad. Sci. (USA)* **90**, 6125–6129.

Burke, C. J., Sanyal, G., Bruner, M. W., Ryan, J. A., LaFemina, R. L., Robbins, H. L., Zeft, A. S., Middaugh, C. R., and Cordingley, M. G. (1992). Structural implications of spectroscopic characterization of a putative zinc finger peptide from HIV-1 integrase. *J. Biol. Chem.* **267**, 9639–9644.

Bushman, F. D. (1994). Tethering human immunodeficiency virus 1 integrase to a DNA site directs integration to nearby sequences. *Proc. Natl. Acad. Sci. (USA)* **91**, 9233–9237.

Bushman, F. D., and Craigie, R. (1991). Activities of human immunodeficiency virus (HIV) integration protein in vitro: Specific cleavage and integration of HIV DNA. *Proc. Natl. Acad. Sci. (USA)* **88**, 1339–1343.

Bushman, F. D., Engelman, A., Palmer, I., Wingfield, P., and Craigie, R. (1993). Domains of the integrase protein of human immunodeficiency virus type 1 responsible for polynucleotidyl transfer and zinc binding. *Proc. Natl. Acad. Sci. (USA)* **90**, 3428–3432.

Bushman, F. D., Fujiwara, T., and Craigie, R. (1990). Retroviral DNA integration directed by HIV integration protein in vitro. *Science* **249**, 1555–1558.

Bushman, F. D., and Miller, M. D. (1997). Tethering human immunodeficiency virus type 1 preintegration complexes to target DNA promotes integration at nearby sites. *J. Virol.* **71,** 458–464.

Bushman, F. D., and Wang, B. (1994). Rous sarcoma virus integrase protein: Mapping functions for catalysis and substrate binding. *J. Virol.* **68,** 2215–2223.

Cai, M., Zheng, R., Caffrey, M., Craigie, R., Clore, G. M., and Gronenborn, A. M. (1997). Solution structure of the N-terminal zinc binding domain of HIV-1 integrase. *Nature Struct. Biol.* **4,** 567–577.

Carteau, S., Hoffmann, C., and Bushman, F. (1998). Chromosome structure and human immunodeficiency virus type1 cDNA integration: centromeric alphoid repeats are a disfavored target. *J. Virol.* **72,** 4005–4014.

Chalker, D. L., and Sandmeyer, S. B. (1990). Transfer RNA genes are genomic targets for *de novo* transposition of the yeast retrotransposon Ty3. *Genetics* **126,** 837–850.

Chalker, D. L., and Sandmeyer, S. B. (1992). Ty3 integrates within the region of RNA polymerase III transcription initiation. *Genes Dev.* **6,** 117–128.

Chen, H., and Engelman, A. (1998). The barrier-to-autointegration protein is a host factor for HIV type 1 integration. *Proc. Natl. Acad. Sci. (USA)* **95,** 15270–15274.

Chen, H., Wei, S.-Q., and Engelman, A. (1999). Multiple integrase functions are required to form the native structure of the human immunodeficiency virus type 1 intasome. *J. Biol. Chem.* **274,** 17358–17364.

Chow, S. A. (1997). In vitro assays for activities of retroviral integrase. *Methods: A Companion to Methods Enzymol.* **12,** 306–317.

Chow, S. A., Vincent, K. A., Ellison, V., and Brown, P. O. (1992). Reversal of integration and DNA splicing mediated by integrase of human immunodeficiency virus. *Science* **255,** 723–726.

Craigie, R., Fujiwara, T., and Bushman, F. (1990). The IN protein of Moloney murine leukemia virus processes the viral DNA ends and accomplishes their integration in vitro. *Cell* **62,** 829–837.

Daniel, R., Katz, R. A., and Skalka, A. M. (1999). A role for DNA-PK in retroviral DNA integration. *Science* **284,** 644–647.

Devine, S. E., and Boeke, J. D. (1996). Regionally specific, targeted integration of the yeast retrotransposon Ty1 upstream of genes transcribed by RNA polymerase III. *Genes Dev.* **10,** 620–633.

Dhar, R., McClements, W. L., Enquist, L. W., and Vande Woude, G. F. (1980). Nucleotide sequences of integrated Moloney sarcoma provirus long terminal repeats and their host and viral junctions. *Proc. Natl. Acad. Sci. (USA)* **77,** 3937–3941.

Donzella, G. A., Leon, O., and Roth, M. (1998). Implication of a central cysteine residue and the HHCC domain of Moloney murine leukemia virus integrase protein in functional multimerization. *J. Virol.* **72,** 1691–1698.

Drelich, M., Wilhelm, R., and Mous, J. (1992). Identification of amino acid residues critical for endonuclease and integration activities of HIV-1 IN protein *in vitro. Virology* **188,** 459–468.

Dyda, F., Hickman, A. B., Jenkins, T. M., Engelman, A., Craigie, R., and Davies, D. R. (1995). Crystal structure of the catalytic domain of HIV-1 integrase: Similarity to other polynucleotidyl transferases. *Science* **266,** 1981–1986.

Eijkelenboom, A. P. A. M., Puras Lutzke, R. A., Boelens, R., Plasterk, R. H. A., Kaptein, R., and Hard, K. (1995). The DNA-binding domain of HIV-1 integrase has an SH3-like fold. *Nature Struct. Biol.* **2,** 807–810.

Eijkelenboom, A. P. A. M., van den Ent, F. M. I., Vos, A., Doreleijers, J. F., Hard, K., Tullius, T. D., Plasterk, R. H. A., Kaptein, R., and Boelens, R. (1997). The solution structure of the aminoterminal HHCC domain of HIV-2 integrase: A three-helix bundle stabilized by zinc. *Curr. Biol.* **7,** 739–746.

Ellison, V., and Brown, P. O. (1994). A stable complex between integrase and viral DNA ends mediates HIV integration in vitro. *Proc. Natl. Acad. Sci. (USA)* **91,** 7316–7320.

Ellison, V., Gerton, J., Vincent, K. A., and Brown, P. O. (1995). An essential interaction between distinct domains of HIV-1 integrase mediates assembly of the active multimer. *J. Biol. Chem.* **270,** 3320–3326.

Engelman, A., Bushman, F. D., and Craigie, R. (1993). Identification of discrete functional domains of HIV-1 integrase and their organization within an active multimeric complex. *EMBO J.* **12**, 3269–3275.

Engelman, A., and Craigie, R. (1992). Identification of conserved amino acid residues critical for human immunodeficiency virus type 1 integrase function in vitro. *J. Virol.* **66**, 6361–6369.

Engelman, A., Englund, G., Orenstein, J. M., Martin, M. A., and Craigie, R. (1995). Multiple effects of mutants in human immunodeficiency virus type 1 integrase on viral replication. *J. Virol.* **69**, 2729–2736.

Engelman, A., Hickman, A. B., and Craigie, R. (1994). The core and carboxy-terminal domains of the integrase protein of human immunodeficiency virus type 1 each contribute to nonspecific DNA binding. *J. Virol.* **68**, 5911–5917.

Engelman, A., Mizuuchi, K., and Craigie, R. (1991). HIV-1 DNA integration: mechanism of viral DNA cleavage and DNA strand transfer. *Cell* **67**, 1211–1221.

Esposito, D., and Craigie, R. (1999). HIV integrase structure and function. *Adv. Virus Res.* **52**, 319–333.

Esposito, D., and Craigie, R. (1998). Sequence specificity of viral end DNA binding by HIV-1 integrase reveals critical regions for protein-DNA interactions. *EMBO J.* **17**, 5832–5843.

Farnet, C. M., and Bushman, F. D. (1997). HIV-1 cDNA integration: Requirement of HMG I(Y) protein for function of preintegration complexes in vitro. *Cell* **88**, 483–492.

Farnet, C. M., and Haseltine, W. A. (1991). Determination of viral proteins present in the human immunodeficiency virus type 1 preintegration complex. *J. Virol.* **65**, 1910–1915.

Farnet, C. M., and Haseltine, W. A. (1990). Integration of human immunodeficiency virus type 1 DNA in vitro. *Proc. Natl. Acad. Sci. (USA)* **87**, 4164–4168.

Fitzgerald, M. L., and Grandgenett, D. P. (1994). Retroviral integration: In vitro host site selection by avian integrase. *J. Virol.* **68**, 4314–4321.

Fletcher, T. M., III., Soares, M. A., McPhearson, S., Huxiong, H., Wiskerchen, M., Muesing, M., Shaw, G. M., Leavitt, A. D., Boeke, J. D., and Hahn, B. H. (1997). Complementation of integrase function in HIV-1 virions. *EMBO J.* **16**, 5123–5138.

Fujiwara, T., and Mizuuchi, K. (1988). Retroviral DNA integration: Structure of an integration intermediate. *Cell* **54**, 497–504.

Gallay, P., Swingler, S., Song, J., Bushman, F., and Trono, D. (1995). HIV nuclear import is governed by the phosphotyrosine-mediated binding of matrix to the core domain of integrase. *Cell* **17**, 569–576.

Gaur, M., and Leavitt, A. D. (1998). Mutations in the human immunodeficiency virus type 1 integrase D,D(35)E motif do not eliminate provirus formation. *J. Virol.* **72**, 4678–4685.

Gerton, J. L., and Brown, P. O. (1997). The core domain of HIV-1 integrase recognizes key features of its DNA substrates. *J. Biol. Chem.* **272**, 25809–25815.

Gerton, J. L., Herschlag, D., and Brown, P. O. (1999). Stereospecificity of reactions catalyzed by HIV-1 integrase. *J. Biol. Chem.* **274**, 33480–33487.

Gerton, J. L., Ohgi, S., Olsen, M., Derisi, J., and Brown, P. O. (1998). Effects of mutations in residues near the active site of human immunodeficiency virus type 1 integrase on specific enzyme-substrate interactions. *J. Virol.* **72**, 5046–5055.

Goff, S. P. (1992). Genetics of retroviral integration. *Annu. Rev. Genet.* **26**, 527–544.

Goodarzi, G., Chiu, R., Brackmann, K., Kohn, K., Pommier, Y., and Grandgenett, D. P. (1997). Host site selection for concerted integration by human immunodeficiency virus type-1 virions in vitro. *Virology* **231**, 210–217.

Goodsell, D. S., Kopka, M. L., Cascio, D., and Dickerson, R. E. (1993). Crystal structure of CATG-GCCATG and its implications for A-tract bending models. *Proc. Natl. Acad. Sci. (USA)* **90**, 2930–2934.

Goulaouic, H., and Chow, S. A. (1996). Directed integration of viral DNA mediated by fusion proteins consisting of human immunodeficiency virus type 1 integrase and *Escherichia coli* LexA protein. *J. Virol.* **70**, 37–46.

Heuer, T. S., and Brown, P. O. (1997). Mapping features of HIV-1 integrase near selected sites on viral and target DNA molecules in an active enzyme-DNA complex by photo-cross-linking. *Biochemistry* **36,** 10655–10665.

Heuer, T. S., and Brown, P. O. (1998). Photo-cross-linking studies suggest a model for the architecture of an active human immunodeficiency virus type 1 integrase-DNA complex. *Biochemistry* **37,** 6667–6678.

Howard, M. T., and Griffith, J. D. (1993). A cluster of strong topoisomerase II cleavage sites is located near an integrated human immunodeficiency virus. *J. Mol. Biol.* **232,** 1060–1068.

Hughes, S. H., Mutschler, A., Bishop, J. M, and Varmus, H. E. (1981). A Rous sarcoma virus provirus is flanked by short direct repeats of a cellular DNA sequence present in only one copy prior to integration. *Proc. Natl. Acad. Sci. (USA)* **78,** 4299–4303.

Hughes, S. H., Shank, P. R., Spector, D. H., Kung, H.-J., Bishop, J. M., and Varmus, H. E. (1978). Proviruses of avian sarcoma virus are terminally redundant, co-extensive with unintegrated linear DNA and integrated into many sites. *Cell* **15,** 1397–1410.

Jenkins, T. M, Engelman, A., Ghirlando, R., and Craigie, R. (1996). A soluble active mutant of HIV-1 integrase: Involvement of both the core and carboxyl-terminal domains in multimerization. *J. Biol. Chem.* **271,** 7712–7718.

Jenkins, T. M., Esposito, D., Engelman, A., and Craigie, R. (1997). Critical contacts between HIV-1 integrase and viral DNA identified by structure-based analysis and photo-crosslinking. *EMBO J.* **16,** 6849–6859.

Ji, H., Moore, D. P., Blomberg, M. A., Braiterman, L. T., Voytas, D. F., Natsoulis, G., and Boeke, J. D. (1993). Hotspots for unselected Ty1 transposition events on yeast chromosome III are near tRNA genes and LTR sequences. *Cell* **73,** 1007–1018.

Johnson, M. S., McClure, M. A., Feng, D.-F., Gray, J., and Doolittle, R. F. (1986). Computer analysis of retroviral pol genes: Assignment of enzymatic functions to specific sequences and homologies with nonviral enzymes. *Proc. Natl. Acad. Sci. (USA)* **83,** 7648–7652.

Jones, K. S., Coleman, J., Merkel, G. W., Laue, T. M., and Skalka, A. M. (1992). Retroviral integrase functions as a multimer and can turn over catalytically. *J. Biol. Chem.* **267,** 16037–16040.

Jonsson, C. B., Donzella, G. A., Gaucan, E., Smith, C. M., and Roth, M. (1996). Functional domains of Moloney murine leukemia virus integrase defined by mutation and complementation analysis. *J. Virol.* **70,** 4585–4597.

Kahn, E., Mack, J. P. G., Katz, R. A., Kulkosky, J., and Skalka, A. M. (1991). Retroviral integrase domains: DNA binding and recognition of LTR sequences. *Nucleic Acids Res.* **19,** 851–860.

Kalpana, G. V., Marmon, S., Wang, W., Crabtree, G. R., and Goff, S. P. (1994). Binding and stimulation of HIV-1 integrase by a human homology of yeast transcription factor SNF5. *Science* **266,** 2002–2006.

Katz, R. A., Gravuer, K., and Skalka, A. M. (1998). A preferred target DNA structure for retroviral integrase in vitro. *J. Biol. Chem.* **273,** 24190–24195.

Katz, R. A., Merkel, G., Kulkosky, J., Leis, J., and Skalka, A. M. (1990). The avian retroviral IN protein is both necessary and sufficient for integrative recombination in vitro. *Cell* **63,** 87–95.

Katz, R. A., Merkel, G., and Skalka, A. M. (1996). Targeting of retroviral integrase by fusion to a heterologous DNA binding domain: *In vitro* activities and incorporation of a fusion protein into viral particles. *Virology* **217,** 178–190.

Katz, R. A., and Skalka, A. M. (1994). The retroviral enzymes. *Annu. Rev. Biochem.* **63,** 133–173.

Katzman, M., Katz, R. A., Skalka, A. M., and Leis, J. (1989). The avian retroviral integration protein cleaves the terminal sequences of linear viral DNA at the in vivo sites of integration. *J. Virol.* **63,** 5319–5327.

Katzman, M., and Sudol, M. (1995). Mapping domains of retroviral integrase responsible for viral DNA specificity and target site selection by analysis of chimeras between human immunodeficiency virus type 1 and visna virus integrases. *J. Virol.* **69,** 5687–5696.

Katzman, M., and Sudol, M. (1998). Mapping viral DNA specificity to the central region of integrase by using functional human immunodeficiency virus type 1/visna virus chimeric proteins. *J. Virol.* **72**, 1744–1753.

Kim, J.-S., and Pabo, C. O. (1998). Getting a handhold on DNA: Design of poly-zinc finger proteins with femtomolar dissociation constants. *Proc. Natl. Acad. Sci. (USA)* **95**, 2812–2817.

Kim, J. M., Vanguri, S., Boeke, J. D., Gabriel, A., and Voytas, D. F. (1998). Transposable elements and genome organization: A comprehensive survey of retrotransposons revealed by the complete *Saccharomyces cerevisiae* genome sequence. *Genome Res.* **8**, 464–478.

Kinsey, P. T., and Sandmeyer, S. B. (1991). Adjacent pol II and pol III promoters: Transcription of the yeast retrotransposon Ty3 and a target tRNA gene. *Nucleic Acids Res.* **19**, 1317–1324.

Kirchner, J., Connolly, C. M., and Sandmeyer, S. B. (1995). Requirement of RNA polymerase III transcription factors for in vitro position-specific integration of a retroviruslike element. *Science* **267**, 1488–1491.

Kitamura, Y., Lee, Y. M. H., and Coffin, J. M. (1992). Nonrandom integration of retroviral DNA *in vitro*: Effect of CpG methylation. *Proc. Natl. Acad. Sci. (USA)* **89**, 5532–5536.

Kondo, E., and Gottlinger, H. G. (1996). A conserved LXXLF sequence is the major determinant in p6^{gag} required for the incorporation of human immunodeficiency virus type 1 Vpr. *J. Virol.* **70**, 159–164.

Kulkosky, J., Jones, K. S., Katz, R. A., Mack, J. P. G., and Skalka, A. M. (1992). Residues critical for retroviral integrative recombination in a region that is highly conserved among retroviral/retrotransposon integrases and bacterial insertion sequence transposases. *Mol. Cell. Biol.* **12**, 2331–2338.

LaFemina, R. L., Callahan, P. L., and Cordingley, M. G. (1991). Substrate specificity of recombinant human immunodeficiency virus integrase protein. *J. Virol.* **65**, 5624–5630.

Larminie, C. G. C., Cairns, C. A., Mital, R., Martin, K., Kouzarides, T., Jackson, S. P., and White, R. J. (1997). Mechanistic analysis of RNA polymerase III regulation by the retinoblastoma protein. *EMBO J.* **16**, 2061–2071.

Laurent, B. C., and Carlson, M. (1992). Yeast SNF2/SWI2, SNF5, and SNF6 proteins function coordinately with the gene-specific transcriptional activators GAL4 and Bicoid [published erratum appears in *Genes Dev.* (1992) **6**, 2233]. *Genes Dev.* **6**, 1707–1715.

Laurent, B. C., Treitel, M. A., and Carlson, M. (1990). The SNF5 protein of *Saccharomyces cerevisiae* is a glutamine- and proline-rich transcriptional activator that affects expression of a broad spectrum of genes. *Mol. Cell Biol.* **10**, 5616–5625.

Leavitt, A. D., Robles, G., Alesandro, N., and Varmus, H. E. (1996). Human immunodeficiency virus type 1 integrase mutants retain in vitro integrase activity yet fail to integrate viral DNA efficiently during infection. *J. Virol.* **70**, 721–728.

Leavitt, A. D., Rose, R. B., and Varmus, H. E. (1992). Both substrate and target oligonucleotide sequences affect *in vitro* integration mediated by human immunodeficiency virus type 1 integrase protein produced in *Saccharomyces cerevisiae*. *J. Virol.* **66**, 2359–2368.

Lee, M. S., and Craigie, R. (1994). Protection of retroviral DNA from autointegration: Involvement of a cellular factor. *Proc. Natl. Acad. Sci. (USA)* **91**, 9823–9827.

Lee, M. S., and Craigie, R. (1998). A previously unidentified host protein protects retroviral DNA from autointegration. *Proc. Natl. Acad. Sci. (USA)* **95**, 1528–1533.

Lee, S. P., and Han, M. K. (1996). Zinc stimulates Mg^{2+}-dependent 3′-processing activity of human immunodeficiency virus type 1 integrase in vitro. *Biochemistry* **35**, 3837–3844.

Lee, S. P., Xiao, J., Knutson, J. R., Lewis, M. S., and Han, M. K. (1997). Zn^{2+} promotes the self-association of human immunodeficiency virus type-1 integrase in vitro. *Biochemistry* **36**, 173–180.

Li, L., Farnet, C. M., Anderson, W. F., and Bushman, F. D. (1998). Modulation of activity of Moloney murine leukemia virus preintegration complexes by host factors in vitro. *J. Virol.* **72**, 2125–2131.

Liu, H., Wu, X., Xiao, H., Conway, J. A., and Kappes, J. C. (1997a). Incorporation of functional human immunodeficiency virus type 1 integrase into virions independent of the gag-pol precursor protein. *J. Virol.* **71,** 7704–7710.

Liu, H., Wu, X., Xiao, H., and Kappes, J. C. (1999). Targeting human immunodeficiency virus (HIV) type 2 integrase protein into HIV type 1. *J. Virol.* **73,** 8831–8836.

Liu, Q., Segal, D. J., Ghiara, J. B., and Barbas, C. F., III. (1997b). Design of polydactyl zinc-finger proteins for unique addressing within complex genomes. *Proc. Natl. Acad. Sci. (USA)* **94,** 5525–5530.

Lodi, P. J., Ernst, J. A., Kuszewski, J., Hickman, A. B., Engelman, A., Craigie, R., Clore, G. M., and Gronenborn, A. M. (1995). Solution structure of the DNA binding domain of HIV-1 integrase *Biochemistry* **34,** 9826–9833.

Love, J. J., Li, X., Case, D. A., Giese, K., Grosschedle, R., and Wright, P. E. (1995). Structural basis for DNA bending by the architectural transcription factor LEF-1. *Nature* **376,** 791–795.

Majors, J. E., and Varmus, H. E. (1981). Nucleotide sequences at host-proviral junctions for mouse mammary tumour virus. *Nature* **289,** 253–258.

Masuda, T., Kuroda, M. J., and Harada, S. (1998). Specific and independent recognition of U3 and U5 *att* sites by human immunodeficiency type 1 integrase *in vivo. J.Virol.* **72,** 8396–8402.

Masuda, T., Planelles, V., Krogstad, P., and Chen, I. S. Y (1995). Genetic analysis of human immunodeficiency type 1 integrase and the U3 *att* site: Unusual phenotype of mutants in the zinc finger-like domain. *J. Virol.* **69,** 6687–6696.

Miller, M. D., Farnet, C. M., and Bushman, F. D. (1997). Human immunodeficiency virus type 1 preintegration complexes: Studies of organization and composition. *J. Virol.* **71,** 5382–5390.

Miller. M. D., Wang, B., and Bushman, F. (1995). Human immunodeficiency virus type 1 preintegration complexes containing discontinuous plus strand are competent to integrate *in vitro. J. Virol.* **69,** 3938–3944.

Mizuuchi, K. (1992). Polynucleotidyl transfer reactions in transpositional DNA recombination. *J. Biol. Chem.* **267,** 21273–21276.

Muller, H. P., and Varmus, H. E. (1994). DNA bending creates favored sites for retroviral integration: An explanation for preferred insertion sites in nucleosomes. *EMBO J.* **13,** 4707–4714.

Mumm, S. R., and Grandgenett, D. P. (1991). Defining nucleic acid-binding properties of avian retrovirus integrase by deletion analysis. *J. Virol.* **65,** 1160–1167.

Pavletich, N. P., and Pabo, C. O. (1993). Crystal structure of a five-finger GLI-DNA complex: New perspectives on zinc fingers. *Science* **261,** 1701–1707.

Petit, C., Schwartz, O., and Mammano, F. (1999). Oligomerization within virions and subcellular localization of human immunodeficiency virus type 1 integrase. *J. Virol.* **73,** 5079–5088.

Pruss, D., Bushman, F. D., and Wolffe, A. P. (1994a). Human immunodeficiency virus integrase directs integration to sites of severe DNA distortion within the nucleosome core. *Proc. Natl. Acad. Sci. (USA)* **91,** 5913–5917.

Pruss, D., Reeves, R.,Bushman, F. D., and Wolffe, A. P. (1994b). The influence of DNA and nucleosome structure on integration events directed by HIV integrase. *J. Biol. Chem.* **269,** 25031–25041.

Pryciak, P. M., Muller, H. P., and Varmus, H. E. (1992a). Simian virus 40 minichromosomes as targets for retroviral integration *in vivo. Proc. Natl. Acad. Sci. (USA)* **89,** 9237–9241.

Pryciak, P. M., Sil, A., and Varmus, H. E. (1992b). Retroviral integration into minichromosomes *in vitro. EMBO J.* **11,** 291–303.

Pryciak, P. M., and Varmus, H. E. (1992). Nucleosomes, DNA-binding proteins, and DNA sequence modulate retroviral integration target site selection. *Cell* **69,** 769–780.

Puras Lutzke, R. A., and Plasterk, R. H. A. (1998). Structure-based mutational analysis of the C-terminal DNA-binding domain of human immunodeficiency virus type 1 integrase: Critical residues for protein oligomerization and DNA binding. *J. Virol.* **72,** 4841–4848.

Puras Lutzke, R. A., Vink, C., and Plasterk, R. H. A. (1994). Characterization of the minimal DNA-binding domain of the HIV integrase protein. *Nucleic Acids Res.* **22,** 4125–4131.

Purcell, D. F. J., and Martin, M. A. (1993). Alternative splicing of human immunodeficiency virus type 1 mRNA modulates viral protein expression, replication, and infectivity. *J. Virol.* **67,** 6365–6378.

Rebar, E. J., and Pabo, C. O. (1994). Zinc finger phage: Affinity selection of fingers with new DNA-binding specificities. *Science* **263,** 671–673.

Reicin, A. S., Kalpana, G., Paik, S., Marmon, S., and Goff, S. (1995). Sequences in the human immunodeficiency virus type 1 U3 region required for *in vivo* and *in vitro* integration. *J. Virol.* **69,** 5904–5907.

Rice, P., Craigie, R., and Davies, D. R. (1996). Retroviral integrases and their cousins. *Curr. Opin. Struct. Biol.* **6,** 76–83.

Richmond, T. J., Finch, J. T., Rushton, B., Rhodes, D., and Klug, A. (1984). Structure of the nucleosome core particle at 7 Å resolution. *Nature* **311,** 532–537.

Roe, T., Chow, S. A., and Brown, P. O. (1997). 3'-End processing and kinetics of 5'-end joining during retroviral integration *in vitro*. *J. Virol.* **71,** 1334–1340.

Rohdewohld, H., Weiher, H., Reik, W., Jaenisch, R., and Breindl, M. (1987). Retrovirus integration and chromatin structure: Moloney murine leukemia proviral integration sites map near DNase I-hypersensitive sites. *J. Virol.* **61,** 336–343.

Satchwell, S. C., Drew, H. R., and Travers, A. A. (1986). Sequence periodicities in chicken nucleosome core DNA. *J. Mol. Biol.* **191,** 659–675.

Scherdin, U., Rhodes, K., and Breindl, M. (1990). Transcriptionally active genome regions are preferred targets for retrovirus integration. *J. Virol.* **64,** 907–912.

Scottoline, B. P., Chow, S., Ellison, V., and Brown, P. O. (1997). Disruption of terminal base pairs of retroviral DNA during integration. *Genes Dev.* **11,** 371–382.

Sherman, P. A., Dickson, M. L., and Fyfe, J. A. (1992). Human immunodeficiency virus type 1 integration protein: DNA sequence requirements for cleaving and joining reactions. *J. Virol.* **66,** 3593–3601.

Shibagaki, Y., and Chow, S. A. (1997). Central core domain of retroviral integrase is responsible for target site selection. *J. Biol. Chem.* **272,** 8361–8369.

Shibagaki, Y., Holmes, M. L., Appa, R. S., and Chow, S. A. (1997). Characterization of feline immunodeficiency virus integrase and analysis of functional domains. *Virology* **230,** 1–10.

Shih, C.-C., Stoye, J. P., and Coffin, J. M. (1988). Highly preferred targets for retrovirus integration. *Cell* **53,** 531–537.

Shimotohno, K., Mizutani, S., and Temin, H. M. (1980). Sequence of retrovirus provirus resembles that of bacterial transposable elements. *Nature* **285,** 550–554.

Stevens, S. W., and Griffith, J. D. (1994). Human immunodeficiency virus type 1 may preferentially integrate into chromatin occupied by L1Hs repetitive elements. *Proc. Natl. Acad. Sci. (USA)* **91,** 5557–5561.

Stevens, S. W., and Griffith, J. D. (1996). Sequence analysis of the human DNA flanking sites of human immunodeficiency virus type 1 integration. *J. Virol.* **70,** 6459–6462.

van den Ent, F. M. I., Vink, C., and Plasterk, R. H. A. (1994). DNA substrate requirements for different activities of the human immunodeficiency virus type 1 integrase protein. *J. Virol.* **68,** 7825–7832.

van den Ent, F. M. I., Vos, A., and Plasterk, R. H. A. (1999). Dissecting the role of the N-terminal domain of human immunodeficiency virus integrase by *trans*-complementation analysis. *J. Virol.* **73,** 3176–3183.

van Gent, D., Vink, C., Oude Groeneger, A. A. M., and Plasterk, R. H. A. (1993). Complementation between HIV integrase proteins mutated in different domains. *EMBO J.* **12,** 3261–3267.

Vijaya, S., Steffen, D. L., and Robinson, H. L. (1986). Acceptor sites for retroviral integrations map near DNase I-hypersensitive sites in chromatin. *J. Virol.* **60,** 683–692.

Vincent, K. A., Ellison, V., Chow, S. A., and Brown, P. O. (1993). Characterization of human immunodeficiency virus type 1 integrase expressed in *Escherichia coli* and analysis of variants with amino-terminal mutations. *J. Virol.* **67**, 425–437.

Vink, C., Oude Groeneger, A. A. M., and Plasterk, R. H. A. (1993). Identification of the catalytic and DNA-binding region of the human immunodeficiency virus type 1 integrase protein. *Nucleic Acids Res.* **21**, 1419–1425.

Vink, C., and Plasterk, R. H. A. (1993). The human immunodeficiency virus integrase protein. *Trends Genet.* **9**, 433–437.

Vink, C., Puras Lutzke, R. A., and Plasterk, R. H. A. (1994). Formation of a stable complex between the human immunodeficiency virus integrase protein and viral DNA. *Nucleic Acids Res.* **22**, 4103–4110.

Vink, C., van Gent, D. C., Elgersma, Y., and Plasterk, R. H. A. (1991). Human immunodeficiency virus integrase protein requires a subterminal position of its viral DNA recognition sequence for efficient cleavage. *J. Virol.* **65**, 4636–4644.

Wang, B. S., and Pabo, C. O. (1999). Dimerization of zinc fingers mediated by peptides evolved *in vitro* from random sequences. *Proc. Natl. Acad. Sci. (USA)* **96**, 9568–9573.

Wei, S.-Q., Mizuuchi, K., and Craigie, R. (1997). A large nucleoprotein assembly at the ends of the viral DNA mediates retroviral DNA integration. *EMBO J.* **16**, 7511–7520.

Whitcomb, J. M., and Hughes, S. H. (1992). Retroviral reverse transcription and integration: Progress and problems. *Annu. Rev. Cell Biol.* **8**, 275–306.

Withers-Ward, E. S., Kitamura, Y., Barnes, J. P., and Coffin, J. M. (1994). Distribution of targets for avian retrovirus DNA integration *in vivo. Genes Dev.* **8**, 1473–1487.

Wittig, B., Dorbic, T., and Rich, A. (1991). Transcription is associated with Z-DNA formation in metabolically active permeabilized mammalian cell nuclei [published erratum appears in *Proc. Natl. Acad. Sci. (USA)* (1991) **88**, 6898]. *Proc. Natl. Acad. Sci. (USA)* **88**, 2259–2263.

Wittig, B., Wölfl, S., Dorbic, T., Vahrson, W., and Rich, A. (1992). Transcription of human *c-myc* in permeabilized nuclei is associated with formation of Z-DNA in three discrete regions of the gene. *EMBO J.* **11**, 4653–4663.

Wlodawar, A. (1999). Crystal structures of catalytic core domains of retroviral integrases and role of divalent cations in enzymatic activity. *Adv. Viral Res.* **52**, 335–350.

Woerner, A. M., Klutch, M., Levin, J. G., and Markus-Sekura, C. J. (1992). Localization of DNA binding activity of HIV-1 integrase to the C-terminal half of the protein. *AIDS Res. Hum. Retroviruses* **8**, 2433–2437.

Woerner, A. M., and Marcus-Sekura, C. J. (1993). Characterization of a DNA binding domain in the C-terminus of HIV-1 integrase by deletion mutagenesis. *Nucleic Acids Res.* **21**, 3507–3511.

Wu, W., Liu, H., Xiao, H., Conway, J. A., Hehl, E., Kalpana, G. V., Prasad, V., and Kappes, J. C. (1999). Human immunodeficiency virus type 1 integrase protein promotes reverse transcription through specific interactions with the nucleoprotein reverse transcription complex. *J. Virol.* **73**, 2126–2135.

Wu, X., Liu, H., Xiao, H., Conway, J. A., Hunter, E., and Kappes, J. C. (1997). Functional RT and IN incorporated into HIV-1 particles independently of the Gag/Pol precursor protein. *EMBO J.* **16**, 5113–5122.

Wu, X., Liu, H., Xiao, H., Kim, J., Seshaiah, P., Natsoulis, G., Boeke, J. D., Hahn, B. H., and Kappes, J. C. (1995). Targeting foreign proteins to human immunodeficiency virus particles via fusion with vpr and vpx. *J. Virol.* **69**, 3389–3398.

Yang, F., Leon, O., Greenfield, N. J., and Roth, M. J. (1999). Functional interactions of the HHCC domain of Moloney murine leukemia virus integrase revealed by nonoverlapping complementation and zinc-dependent dimerization. *J. Virol.* **73**, 1809–1817.

Yang, S. W., and Nash, H. A. (1994). Specific photocrosslinking of DNA-protein complexes: Identification of contacts between integration host factor and its target DNA. *Proc. Natl. Acad. Sci. (USA)* **91**, 12183–12187.

Zheng, R., Jenkins, T. M., and Craigie, R. (1996). Zinc folds the N-terminal domain of HIV-1 integrase, promotes multimerization, and enhances catalytic activity. *Proc. Natl. Acad. Sci. (USA)* **93,** 13659–13664.

Zou, S., Ke, N., Kim, J. M., and Voytas, D. F. (1996). The *Saccharomyces* retrotransposon Ty5 integrates preferentially into regions of silent chromatin at the telomeres and mating loci. *Genes Dev.* **10,** 634–645.

Zou, S., and Voytas, D. F. (1997). Silent chromatin determines target preference of the retrotransposon Ty5. *Proc. Natl. Acad. Sci. (USA)* **94,** 7412–7416.

3 Xeroderma Pigmentosum and Related Disorders: Defects in DNA Repair and Transcription

Mark Berneburg† and Alan R. Lehmann*
MRC Cell Mutation Unit
University of Sussex
Falmer, Brighton, BN1 9RR, United Kingdom

I. Introduction
II. Xeroderma Pigmentosum
    A. XPC and Damage Detection
    B. XPA and Damage Verification
    C. XPE: An Inducible Protein Involved in the Recognition of Cpds
    D. XPB: A Helicase for NER and Transcription
    E. XPD: A Helicase Only for NER?
    F. XPG: A Nuclease Also Involved in Repair of Oxidative Damage
    G. XPF: Nuclease for 5′ Incision
    H. XP Variant: A DNA Polymerase Involved in Trans-Lesion Synthesis
III. Cockayne Syndrome
    A. Clinical and Cellular Characteristics
    B. CS and Transcription
    C. Oxidative Damage
IV. Trichothiodystrophy
V. Outstanding Clinical Questions
VI. Concluding Remarks
    References

† Present address: Department of Dermatology, Heinrich Heine University, D-40225 Düsseldorf, Germany
*Corresponding author: Telephone: 44-12-7367-8120. Fax 44-12-7367-8121. E-mail: a.r.lehmann @sussex.ac.uk

ABSTRACT

The genetic disorders xeroderma pigmentosum (XP), Cockayne syndrome (CS), and trichothiodystrophy (TTD) are all associated with defects in nucleotide excision repair (NER) of DNA damage. Their clinical features are very different, however, XP being a highly cancer-prone skin disorder, whereas CS and TTD are cancer-free multisystem disorders. All three are genetically complex, with at least eight complementation groups for XP (XP-A to -G and variant), five for CS (CS-A, CS-B, XP-B, XP-D, and XP-G), and three for TTD (XP-B, XP-D, and TTD-A). With the exception of the variant, the products of the XP genes are proteins involved in the different steps of NER, and comprise three damage-recognition proteins, two helicases, and two nucleases. The two helicases, XPB and XPD, are components of the basal transcription factor TFIIH, which has a dual role in NER and initiation of transcription. Different mutations in these genes can affect NER and transcription differentially, and this accounts for the different clinical phenotypes. Mutations resulting in defective repair without affecting transcription result in XP, whereas if transcription is also affected, TTD is the outcome. CS proteins are only involved in transcription-coupled repair, a subpathway of NER in which damage in the transcribed strands of active genes is rapidly and preferentially repaired. Current evidence suggests that they also have an important but not essential role in transcription. The variant form of XP is defective in a novel DNA polymerase, which is able to synthesise DNA past UV-damaged sites.   © 2001 Academic Press.

I. INTRODUCTION

About 10 genetic disorders are now known to be associated with defects in DNA repair or other responses to DNA damage. The prototype disease in this category is the highly cancer-prone xeroderma pigmentosum (XP). Most XP patients are deficient in their ability to remove DNA damage induced by ultraviolet (UV) light. This is a result of mutations in one of the genes whose products are involved in the process of nucleotide excision repair (NER), a highly conserved mechanism that repairs a wide range of bulky, helix-distorting lesions in cellular DNA (Friedberg *et al.*, 1995). One of the most important functions of NER is the removal of damage inflicted by UV light from the sun, such as cyclobutane-pyrimidine dimers (CPD) and 6-4, pyrimidine pyrimidone photoproducts (6-4 PP). NER is a tightly regulated process that involves at least 30 gene products (Friedberg *et al.*, 1995) acting in a sequential manner to recognize DNA damage, recruit the repair machinery, open up the damaged site, incise on both sides of the damage, remove the oligonucleotide containing the damaged site, and fill in the gap with normal DNA (Friedberg *et al.*, 1995; Wood, 1996, 1997; de Laat *et al.*, 1999). Two other genetic disorders, Cockayne syndrome (CS) and trichothiodystrophy (TTD), are also associated with defects in NER, but their clinical features are quite different from those of XP (Lehmann, 1995), a major difference being the lack of skin cancers associated with CS and TTD. There are two distinct subpathways of NER: global genome repair (GGR) and transcription-coupled repair (TCR). GGR nonselectively removes damage relatively slowly from the whole genome, whereas TCR is a rapid repair mechanism that removes damage specifically from the transcribed strand of active genes. CS cells are deficient in TCR but have normal GGR (van Hoffen et al., 1993).

TCR is one of the important links between repair and transcription. A further link was revealed with the unexpected discovery that the basal transcription factor TFIIH has a dual role, in NER as well as in transcription (Schaeffer et al., 1993). These connections between NER and transcription are of great importance for our understanding of the complex etiology of XP, CS, and TTD (Hoeijmakers et al., 1996; Lehmann, 1995).

There have been major advances in our understanding of the NER disorders over the last few years, and there are numerous reviews available covering the different aspects of NER and DNA-repair disorders (de Laat et al., 1999; Friedberg, 1996; Friedberg et al., 1995; Taylor and Lehmann, 1998; van Gool et al., 1997b; Wood, 1996, Wood, 1997). We have therefore concentrated in this review on the most recent developments, which have revealed further fascinating and unexpected complexities in these disorders.

II. XERODERMA PIGMENTOSUM

Patients with XP show sun sensitivity, pigmented macules (freckles), achromic spots, telangiectasia, and an approximately 2000-fold increased risk of developing skin cancers, including basal and squamous cell carcinomas as well as malignant melanomas (Bootsma et al., 1998; Kraemer et al., 1987). All the skin abnormalities are located exclusively on sunlight-exposed areas. There is also evidence for an increased prevalence of internal tumors (Kraemer et al., 1984). Some patients with XP exhibit neurological abnormalities. At the cellular level, fibroblasts as well as T-lymphocytes and lymphoblastoid cells from most patients are hypersensitive to the lethal effects of UV-C and UV-B. The sensitivity to UV is caused by a defect in removing bulky, helix-distorting lesions from their DNA. Genetic studies on cultured cells have revealed seven complementation groups (A–G) which are deficient in NER as well as a variant form, in which NER and UV sensitivity are normal. The different complementation groups exhibit specific features distinguishing them from other groups. The underlying genetic and molecular defects have been characterized for most of the complementation groups, and the roles that the corresponding gene products play in the normal processes have in turn helped to understand the pathological processes leading to the specific abnormalities in each complementation group. Therefore, in the following section the clinical and cellular features are correlated with the underlying molecular defects. The genes, their encoded proteins, and the corresponding features associated with affected individuals and laboratory-generated mutant mice are summarized in Table 3.1.

A. XPC and damage detection

1. Cellular studies

XP-C is one of the most frequent XP complementation groups in Europe and the United States, but it is rare in Japan. XP-C patients in general have severe skin symptoms, but rarely show any neurological abnormalities. The sensitivity of XP-C cells to UV-induced cell killing is not as marked as in some of the other groups, although they do exhibit a severely reduced level of NER. XP-C cells are deficient in global genome repair. Their ability to carry out transcription-coupled repair is normal, however (van Hoffen *et al.*, 1995), and RNA synthesis recovers to normal rates after UV irradiation with similar kinetics to normal cells (Mayne and Lehmann, 1982). This probably accounts for their relatively modest sensitivity to the lethal effects of UV light.

2. Role of XPC protein

The role of the XPC protein in NER had been unclear until recent work demonstrated that a complex of XPC and HR23B (one of the two human homologs of the yeast Rad23 NER protein) (Masutani *et al.*, 1994) is involved in the earliest step of NER, namely, the detection of DNA damage (Figure 3.1). This role had been previously ascribed to XPA (see below). Sugasawa *et al.* (1998) showed, however, that XPC-HR23B has a high affinity for DNA and that its affinity for damaged

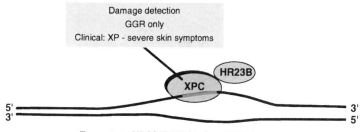

Figure 3.1. XPC/HR23B binds to DNA damage.

Table 3.1. XP and CS Genes, Proteins, and Clinical Features

Complementation group/gene	Frequency	Protein function	Clinical features	Mice
XP-A	Frequent in Japan and the West	Verification of DNA damage	Very severe skin and neurological abnormalities	Viable Extremely susceptible to skin cancer No neurological abnormalities
XP-B	Very rare	3'-to-5' helicase in transcription and NER Subunit of TFIIH	XP in combination with CS: varying severity TTD: mild	
XP-C	Frequent in West	Detection of DNA damage GGR	Severe skin abnormalities	
XP-D	Frequent in West	5'-to-3' helicase Subunit of TFIIH Main function NER, but required for transcription	XP-D: mild to severe/neurological defects TTD: varying symptoms XP/CS: two cases, very severe	Knockout mice not viable TTD mice: hair abnormalities, premature ageing
XP-E	Rare	Recognition of CPDs in nontranscribing DNA	Mild symptoms	
XP-F	Rare	5' endonuclease	Mild symptoms	
XP-G	Rare	3'-endonuclease Removal of oxidative damage	XP-G: mild symptoms XP/CS: very severe	Postnatal growth failure, premature death
XP variant	Intermediate	Polymerase η synthesis past CPD	Skin abnormalities; varying severity	
CS-A	Rare	Transcription-coupled repair	Growth retardation, neurological disorders, sun sensitivity	
CS-B	Intermediate	Interaction with RNA polymerase II Transcription-coupled repair	Growth retardation, neurological disorders, sun sensitivity	Mice develop skin cancers and show no neurological abnormalities

DNA is at least 10 times higher than for undamaged DNA. Furthermore, us-ing a damage-recognition competition assay, Sugasawa *et al.* found that DNA that was preincubated with an extract lacking the XPA protein but containing XPC protein was always repaired more efficiently than vice versa. This indicates that XPC binds to damaged DNA before the binding of XPA. Thus, Sugasawa *et al.* suggested that the XPC–HR23B complex detects the DNA damage first and binds to it independently from the XPA protein, which is subsequently re-cruited. The notion that XPC–HR23B binds to DNA damage first is in line with the work of Wakasugi and Sancar (1998), who found that a complex of repair proteins was able to bind to damaged DNA. This binding required XPC, but the bound complex did not contain XPC. They therefore proposed that XPC ful-filled the role of a molecular matchmaker. This is also in accord with the work of Li *et al.* (1998), who found, in a cell-free NER system, a long-lived DNA–protein complex involving the XPA protein and the basal transcription factor TFIIH. Binding of this complex to DNA-lesions required XPC protein (Li *et al.*, 1998).

Furthermore, a study by Baxter and Smerdon (1998), employing organ-omercurial chromatography, showed that nucleosome folding occurred transiently during the NER process and that this required the XPC protein. This suggests that the XPC protein might have a role not only in damage detection and recruitment of the XPA protein but that it might also provide the necessary conformation of the DNA–protein complex to facilitate NER initiation. Since XPC is in-volved in GGR but not TCR, it is evident that other proteins must carry out damage recognition in the latter process. It is likely that in transcribing DNA, the stalling of RNA polymerase at the lesion in some way generates the recog-nition signal, and obviates the need for XPC in this subpathway (see Section III.B). The two different repair pathways then come together in the recruitment of XPA (together with RPA and TFIIH) and thereafter NER follows a common pathway. In support of this hypothesis, Mu and Sancar (1997) used a model sub-strate in which a CPD was present in the middle of a "bubble" substrate. Repair of the lesion in this conformation did not require the XPC protein, suggest-ing that this conformation mimicked that of DNA containing a stalled RNA polymerase.

3. Germline mutations in the XPC gene

Mutations have been identified in several XP-C patients (Li *et al.*, 1993; Chavanne et al., 2000), and in nearly all cases they result in protein truncations. This demon-strates that XPC is nonessential for life, which is consistent with the viability of XP-C knockout mice generated in several laboratories (Berg *et al.*, 1998; Cheo *et al.*, 1997; Sands *et al.*, 1995). These mice are healthy, but sensitive to the carcinogenic effects of UV-B irradiation (Cheo *et al.*, 1996; Sands *et al.*, 1995).

4. Mutations in tumors of XP patients

XP individuals have a 2000-fold increase in the incidence of skin tumors. These are almost exclusively confined to areas exposed to sunlight and are one of the consequences of unrepaired damage. Like other cancers, skin tumors result from a series of mutations and other genetic changes in critical oncogenes or tumor suppressor genes. Mutations induced by UV light have a characteristic "signature," the majority being C : G to T : A transitions, of which a proportion are CC-to-TT tandem mutations. The latter type of mutation is produced almost exclusively by UV light. Sarasin and colleagues have examined the status of the *p53* gene in tumors from several XP patients and found that 80% of them contained mutations in the *p53* gene, 89% of which were C-to-T transitions and 61% of these were CC-to-TT tandem mutations. Almost all of the mutations could be attributed to dipyrimidine sites on the nontranscribed strand of the DNA. Most of the patients in this study were in the XP-C group, in which only damage in the transcribed strand of active genes is repaired. These data are consistent with an early stage of tumor formation being a UV light-induced mutation resulting from unrepaired damage on the nontranscribed strand of the *p53* gene (Dumaz *et al.*, 1993; Giglia *et al.*, 1998). Similar results were obtained in an earlier, more limited study by Matsumara *et al.* (1995).

B. XPA and damage verification

1. Function of XPA protein

XP-A is the most common XP in Japan (Takebe *et al.*, 1977) and is also relatively frequent in the West. Patients with XP-A are usually severely affected clinically, with both cutaneous and neurological abnormalities (Figure 3.2). XP-A cells have very low levels of NER and are most sensitive to killing by UV. XP-A cells

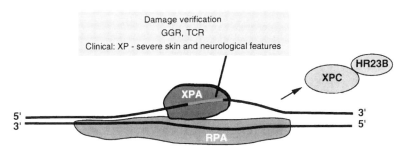

Figure 3.2. XPA/RPA verifies DNA damage.

are defective in both TCR and GGR, which indicates that the XPA protein is involved in both pathways.

Sequence analysis of the XPA gene product has revealed zinc-finger motifs indicating that the XPA protein might bind to DNA (Morita et al., 1996; Tanaka et al., 1990). There has been uncertainty about the number of zinc atoms bound to full-length XPA, since there is a possibility of one or two zinc-binding domains in the C-terminal region of XPA (Hess et al., 1998; Tanaka et al., 1990). However, recent X-ray absorption fine structure analysis and mass spectroscopy of the XPA protein in Xenopus laevis indicate that there is only one zinc-binding motif (Buchko et al., 1999).

The XPA protein binds to DNA and it has a higher affinity to damaged than to undamaged DNA (Asahina et al., 1994; Jones and Wood, 1993), indicating a role for the XPA protein in the early stages of NER. It had been hypothesized for many years that XPA was responsible for the detection of DNA damage. However, although XPA shows an increased affinity to damaged DNA, the fivefold increased affinity of XPA for damaged DNA is much too small for the high lesion specificity observed in NER. In line with this, work by Sugasawa et al. (1998) described above has shown that the XPA protein binds to DNA after the binding of the XPC–HR23B complex. This indicates that although XPA plays a role in the early stages of NER, it is involved in a function after damage recognition, possibly damage verification or recruitment of other NER subunits such as RPA and TFIIH (Nocentini et al., 1997). A role for the XPA protein in damage verification, combined with the damage recognition role of XPC, would explain the damage specificity of NER and, since verification of the existing damage is required in both TCR and GGR, it would also explain the deficiency of XP-A cells in both pathways.

2. XPA mutations

Mutations have been identified in many XP-A individuals, and a very pronounced founder effect has been discovered in the Japanese XP-A population. About 85% of Japanese XP-A patients have the same mutation, namely, a change in the splice acceptor site of intron 3, resulting in a truncated protein that is lacking the C-terminal half (Nishigori et al., 1994; States et al., 1998). Most of the other mutations also result in protein truncations (Cleaver et al., 1999). The few missense mutations that have been identified are in the zinc-finger domain and interfere with DNA binding. All these patients are severely affected. In contrast, some more mildly affected patients are mutated in the final exon 6, whose function appears to be less critical (Nishigori et al., 1993, 1994). The mutation spectrum of XP-A patients suggests that the XPA gene is nonessential for life, and again this is consistent with the viability of the XP-A knockout mice. These mice are healthy. They do not show the neurological abnormalities of the human XP-As.

Like human XP-A patients, they are extremely sensitive to skin carcinogenesis induced by UV-B or dimethylbenzanthracene (de Vries *et al.*, 1995; Nakane *et al.*, 1995).

C. XPE: An inducible protein involved in the recognition of CPDs

Clinically, XP-E is one of the least severe forms of XP, with only mild dermatological manifestations and no neurological abnormalities. At least 19 cases have been reported (Rapic Otrin *et al.*, 1998), and cells from these patients generally have a NER level ranging from 40% to 60% of normal cells and only slight UV sensitivity. A protein that bound specifically to DNA damaged by UV light or cisplatin was found to be absent in an XP-E strain (Chu and Chang, 1988). This has been reproduced in several laboratories, but in an extensive survey this DNA-binding activity was defective in only 3 of 12 cases with XP-E. Damage-binding activity is missing from some XP-E cells designated DDB$^-$ but is present in others (DDB$^+$) (Keeney *et al.*, 1992). This DNA damage-binding activity (UV-DDB) has been purified as a single 127-kDa protein (Abramic *et al.*, 1991) and as a complex with two subunits of 127 and 48 kDa (Keeney *et al.*, 1993). Sequence analysis of the *p48* subunit has shown mutations in the *p48* gene of XP-E patients, and it has been suggested that they are causative for XP-E (Nichols *et al.*, 1996). Work by Rapic Otrin *et al.* (1998) showed that three newly identified XP-E patients were deficient in UV-DDB activity. Work of Hwang *et al.* (1998, 1999) has shown that the *p48 XPE* gene is inducible by UV irradiation and that this induction is dependent on *p53*. This work also showed that XPE is specifically involved in the global genome repair of CPDs, but is dispensable for transcription-coupled repair of CPDs and for all repair of 6-4 PPs. It is likely therefore that XPE plays a specific role in recognizing CPDs in nontranscribing DNA.

D. XPB: A helicase for NER and transcription

1. Dual role of TFIIH in NER and transcription

There are only three families described in the literature in the XP-B complementation group (Robbins *et al.*, 1974; Vermeulen *et al.*, 1994a; Weeda *et al.*, 1997b). Two have the combined features of XP and CS (in one case mild, the other severe); the third has TTD. The first clue to solving this enigma came from the unexpected discovery that the XPB protein is the largest subunit of the basal transcription factor TFIIH (Schaeffer *et al.*, 1993). This subsequently led to the findings that XPD is also a subunit of TFIIH, and that the whole TFIIH complex has a dual role, in NER as well as in transcription. This is the most striking example of several instances in which DNA repair proteins have multiple functions [see below and Lehmann (1998)]. Since the transcriptional function of TFIIH is

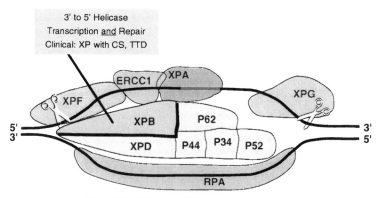

Figure 3.3. XPB in TFIIH opens out damaged site.

vital for life, *XPB* is an essential gene, and the paucity of XP-B families suggests that the transcriptional function of the *XPB* gene is very intolerant of mutations.

The involvement of TFIIH in both NER and transcription implies that mutations in the TFIIH genes may affect NER, transcription, or both. Bootsma and Hoeijmakers (1993) proposed that defects in NER alone result in the XP phenotype, whereas subtle defects in transcription result in CS or TTD, thus making XP a repair deficiency syndrome while CS and TTD are transcription deficiency syndromes. In order to resolve the specific roles of TFIIH components in NER and transcription, various experimental systems have been utilized. Like all the NER genes, XPB is highly conserved, the yeast homolog being designated *RAD25*. The XPB protein has a 3′-to-5′ helicase (see Figure 3.3) (Ma *et al.*, 1994), and DNA-dependent ATPase activity (Drapkin *et al.*, 1994; Guzder *et al.*, 1994; Schaeffer *et al.*, 1993, 1994). The yeast system indicates that the ATPase activity of XPB is required for both transcription and NER, since a mutation in the ATP-binding site of *Rad25* results in a defect in both transcription and NER (Park *et al.*, 1992).

2. Involvement of XPB in opening out structures

The architecture of the TFIIH complex has been investigated in detail and reconstitution of TFIIH subunits including XPB has enabled their roles to be determined at the molecular level. Tirode *et al.* (1999) showed that reconstituted TFIIH contained all its known subunits and that it had the same levels of helicase, ATPase, and transcriptional activities as conventionally purified TFIIH from HeLa cells. In this study, reconstituted protein complexes that lacked the XPB subunit were compared with normal extracts and XPB was found to be essential for the transcription reaction. Further experiments indicated that the role of the XPB helicase in transcription initiation is the opening of the promoter region. This was corroborated and substantiated by a study from the same group in which TFIIH was isolated from cell extracts of two XP-B patients carrying different mutations in the *XPB* gene. These two mutations impaired the helicase, ATPase, and transcriptional activities of the extracts differently (Coin *et al.*, 1999). Furthermore, when the transcription substrate contained an artificially open promoter, the XPB protein was no longer required. This provided strong evidence that the role of XPB in transcription is indeed the opening up of the promoter region. With regard to its function in NER, *in vivo* and *in vitro* studies (Evans *et al.*, 1997b; Vermeulen *et al.*, 1994b; Weeda *et al.*, 1990) showed that the *XPB* mutations described before result in an almost total inhibition (around 95%) of NER, indicating that XPB is indispensable for NER. Furthermore, an intermediate step in the NER process is the formation of open complexes, prior to incision of the DNA on either side of the damage. XPB is required for this open-complex formation, suggesting that within the TFIIH complex, it performs similar roles in NER and transcription. This probably accounts for the rarity of XP-B patients, who must have sufficiently severe mutations to affect repair and give an XP phenotype, yet remain viable by having only a minor effect on transcription.

Evidence for an additional more subtle role of XPB comes from the work of Evans *et al.* (1997b), who analyzed open-complex formation using an *in vitro* NER system. They found that extracts of one mutant XP-B line were unable to form open complexes. In contrast, extracts from another patient with a severe defect in NER did form open complexes, and incised DNA on the 3' side of the damage, but were unable to carry out the subsequent 5' incision (see below for details of incision steps). These results implicate a role for XPB in interactions with other NER components.

In summary, XPB is tightly bound in the core of TFIIH and is principally, but not exclusively, involved in generating open structures for both transcription and NER.

E. XPD: A helicase only for NER?
1. Genotype-phenotype relationships in XPD

The XPD protein is a 5'–3' helicase subunit of TFIIH and as such one might anticipate that the effects of *XPB* and *XPD* mutations would be quite similar. In fact they are quite different. Patients with mutations in the *XPD* gene are relatively frequent (Lehmann, 1995; Taylor *et al.*, 1997). This implies that the transcription function of XPD is much more tolerant of mutations than XPB. XP-D cells are defective in both TCR and GGR and the *XPD* gene is unique, since mutations in it can lead to all three disorders, XP, TTD, and XP with CS. The transcription syndrome hypothesis leads to the prediction that mutations in *XPD* in TTD patients affect transcription, whereas those in XP patients affect only NER.

This further predicts that the sites of the mutations in the *XPD* gene are disease specific. Initial determinations of mutations in *XPD* did not appear to accord with this prediction, since no pattern of disease specificity was discernible and there were several sites where the same mutation was found in both XP and TTD patients. The mutation spectrum is complicated, however, by the fact that many of the individuals are compound heterozygotes (i.e., mutations occur at different sites in the two alleles). If, however, one of the two mutations completely inactivates the protein, the XPD function would have to be carried out by the gene product of the other allele. In the human system this is difficult to resolve. Like XPB, XPD is highly conserved from yeast to man. Taylor *et al.* (1997) took advantage of this by employing a yeast system that allows investigation of mutations in a haploid background and they showed that in many XP-D and TTD cases, one of the mutations was lethal, indicating that in these patients there is only one causative mutation. In particular, they showed that all the sites at which the same mutation had been found in XP and TTD patients were in fact lethal alleles. When the lethal alleles were removed from the mutation spectrum, the remaining causative mutations for the XP-D and TTD patients were indeed distinct from each other, consistent with the idea that the site of the mutation in the *XPD* gene determines the distinct clinical phenotypes of XP-D and TTD.

Further definitive evidence to support this contention has come from the generation of an XP-D mutant mouse. This contained the mutation R722W in the *XPD* gene, a mutation found in four TTD patients. De Boer *et al.* (1998) made the exciting finding that the resulting mouse did indeed have the features of TTD with brittle hair, which went through cyclic periods of loss and regrowth.

Recent work has shown that cells from the two XP-D patients with the combined features of XP and CS respond to UV irradiation in a unique way. The DNA damage appears to generate breaks *in trans* in the DNA at sites distant from the damage (Berneburg *et al.*, 2000b).

2. Role of XPD helicase in TFIIH

More insights into the specific role of XPD and its relationship to clinical features has come both from *in vitro* analysis of TFIIH functions and from analysis of patient mutations. The *XPD* gene encodes a helicase functioning in the 5′-to-3′ direction. In the yeast homolog Rad3, a mutation in the ATP-binding site abolished NER, but did not affect viability (Sung *et al.*, 1988) and the TFIIH of this mutant was active in transcription (Feaver *et al.*, 1993). This showed that, in contrast to XPB, the helicase activity of XPD is required only for NER and not for basal transcription (Figure 3.4). Since many mutations have a drastic effect on NER but only minor effects on transcription, it is likely that the function of XPD in TFIIH is largely structural. As long as the protein is present in the TFIIH complex, it can tolerate substantial alterations to its structure. Detailed *in vitro*

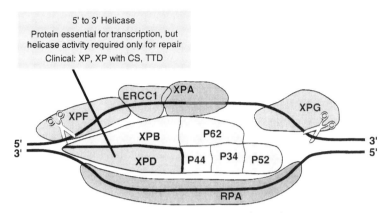

Figure 3.4. XPD in TFIIH opens out damaged site.

studies confirm that XPD is relatively loosely bound in the TFIIH complex. The groups of Reinberg, Sancar, and Egly showed that the core of TFIIH contained five components including XPB, and that XPD acted as a bridge to bind three further components, known as the CAK complex (Drapkin *et al.*, 1996; Reardon *et al.*, 1996; Tirode *et al.*, 1999). XPD and CAK were dispensable for transcription *in vitro*, but when present they had stimulatory effects. This stimulatory effect of XPD was reduced but not abolished in mutants lacking the helicase activity. TFIIH isolated from cells containing the mutation R683W, which is the most common mutation in XP-D patients, contained less XPD, cdk7, and cyclin H protein than wild-type extracts. Even after adjusting the protein amounts to equivalent levels, TFIIH from the XPD-mutated extracts exhibited a weaker helicase and kinase activity than wild-type proteins. However, the 50% decrease in transcription of XPD-defective extracts did not parallel the decrease of approximately 80% in the XPD helicase and kinase activity.

The function of the XPD helicase has also been investigated by Coin *et al.* (1999). They showed that the C-terminus of the XPD subunit interacts with the p44 TFIIH subunit, and that this interaction results in a 10-fold stimulation of the helicase activity. They then immunopurified TFIIH from normal cells and from XP-D cells with different mutations in the *XPD* gene. These mutations included the most frequent XP mutation (R683W), a mutation that gives rise to a combined phenotype of XP with CS (G675R), a 15-bp deletion (716–730) and (R722W), which are both associated with TTD. All of these were located at the C-terminal end of the *XPD* gene. In addition they constructed a mutation in the ATP-binding site of the XPD protein (K48R). TFIIH from cell extracts from the human disorders all showed 5′-to-3′ helicase activity, whereas, as expected, the K48R mutation abolished activity. The co-expression of the patient mutant proteins with p44 showed that these mutations abolished the interaction between XPD and p44, and the helicase activity of XPD was not stimulated by p44. These results explained the effects of mutations in the C-terminal third of the protein on DNA repair activity. However, they did not shed any further light on the disease specificity of these mutations, since they were associated with different disease phenotypes.

The picture emerging from these studies is that it is the helicase activity of XPD within the TFIIH complex that determines the repair capacity of the cell. This is in turn determined by the effect of mutations directly on the active site itself, but also on the interaction with the p44 subunit of TFIIH. The effects on transcription are largely determined by the ability of the mutated XPD protein to maintain the integrity of the TFIIH holo-complex.

F. XPG: A nuclease also involved in repair of oxidative damage

1. Genotype–phenotype relationships in XPG

Xeroderma pigmentosum complementation group G is clinically rather rare, but nevertheless very heterogenous. Nearly all XP-G patients have very low levels of NER. The clinical spectrum ranges from symptoms that are very mild, with little sun sensitivity, to severe, with associated symptoms of CS. Cloning of the gene and analysis of its product showed that it was a structure-specific nuclease, whose reaction specificity made it ideally suited to incising DNA on the 3′ side of the damaged site. With a damaged plasmid substrate, the 3′ incision is carried out by XPG (see Figure 3.5) and the reaction is absolutely dependent on the previous actions of XPA, XPC, and TFIIH (Cloud *et al.*, 1995; Evans *et al.*,

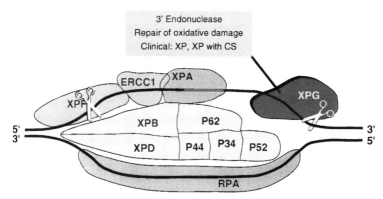

Figure 3.5. XPG cuts 3′ to damage.

1997a; O'Donovan *et al.*, 1994). Sequence analysis of the *XPG* gene has shown that patients with the milder symptoms have missense mutations, which abolish nuclease activity. In contrast, the mutations leading to a combined phenotype of XP-G with CS all result in severely truncated XPG proteins (Nouspikel *et al.*, 1997). This raised the possibility of a second function of the XPG protein, which was retained in the missense mutants even though nuclease activity was abolished. The viability of XP-G patients with severely truncated XPG protein indicates that XPG does not have an essential role. This is confirmed by the viability of XPG knockout mice. These mice were very sick, however, with severe growth retardation and death by 23 days postpartum (Harada *et al.*, 1999).

2. XPG and repair of oxidative damage

The increased severity of the abnormalities in XPG mice compared to XPA mice supports the idea of a second function for the XPG protein. This additional function may be a role in the repair of oxidative damage, and it is possible that a deficiency in this function may be responsible for the features of CS in XP-G patients with the combined features of XP and CS. Cells from XP-G/CS patients are deficient in the transcription coupled repair of thymine glycols, which, in contrast, is found to be normal in XP-A, XP-F, and XP-G cells from patients with XP alone (Cooper *et al.*, 1997). These results indicate that XPG may have a function in TCR of oxidative damage and that a defect in this pathway may contribute to the defects associated with CS symptoms in XP-G/CS patients. Recent *in vitro* studies have provided a possible explanation for this specific deficiency (Klungland *et al.*, 1999). Oxidative damage to DNA is normally removed by base excision repair (BER). Thymines converted into thymine glycol (Tg) by oxidative damage can be removed by BER, the first step being carried out by the human Nth1 DNA-glycosylase, which simultaneously excises the Tg residue and cleaves the DNA on the 3' side of the resulting abasic site. Klungland *et al.* showed that hNth1 activity is stimulated strongly by the XPG protein. A wide range of other DNA-binding proteins including PCNA, RPA, FEN-1, as well as XPA and XPC did not show this stimulatory effect, indicating that it was specific for the XPG protein. The stimulatory effect observed with XPG was furthermore confirmed to be specific for hNth1, since other downstream proteins of BER were not stimulated. XPG appeared to stimulate the binding of hNth1 to DNA. Furthermore, the XPG protein mutated in the nuclease active site maintained its ability to stimulate Nth1. Similar findings were reported by Bessho (1999). These data are consistent with a second role for XPG, in the repair of oxidative damage. Null mutations result in truncations that abolish all functions of XPG, whereas missense mutations can abolish nuclease activity while leaving oxidation repair functions intact. As described above, it has been hypothesized that mutations in the *XPG* gene can cause XPG alone or XP in combination with CS and that a defect of the XPG protein in the repair of oxidative damage may be responsible for the symptoms of CS in this disorder. According to this hypothesis, mutants causing XP with CS should be deficient and mutants causing XP alone should be proficient in the removal of Tg. Unexpectedly, mutants for XPG alone and XP/CS were both as efficient as wild-type protein in Tg removal from the genome overall. The deficiency in the removal of oxidative damage in XPG with CS and in CS (Cooper *et al.*, 1997) appears to be confined to the transcription-coupled repair of these lesions.

Taken together, there is substantial direct and indirect evidence for the involvement of XPG in the repair of oxidized bases and circumstantial evidence that the loss of oxidative damage repair function may play a causative role in the pathogenesis of CS symptoms in the combined disorder XP-G/CS.

G. XPF: Nuclease for 5′ incision

Incision of damaged DNA 5′ to the damaged site is effected by a second structure-specific nuclease, a heterodimer between the products of the *XPF* and *ERCC1* genes (Figure 3.6). XPF/ERCC1 has precisely the opposite polarity from XPG and cuts the DNA 5′ to the damage after XPG has incised on the 3′ side (Sijbers *et al.*, 1996). Mutations have been identified in several XP-F patients and they are all missense mutations. Defects in the partner protein, ERCC1, have not been found in any patients. However, ERCC1 knockout mice have been generated. They are viable but very sick, surviving for only a few weeks. The phenotype of the ERCC1 mice is much more severe than that of XPA mice even though the XPA mice are completely deficient in NER (McWhir *et al.*, 1993; Weeda *et al.*, 1997a). This implies that ERCC1 and, by implication, its partner XPF provide a further example of repair proteins having other roles. In the case of ERCC1/XPF a probable second function is in certain types of recombination. The yeast homologs

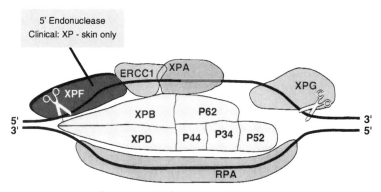

Figure 3.6. XPF/ERCC1 cuts 5′ to damage.

Rad1 and Rad10 are known to be required for recombination between short repeats (Fishman-Lobell and Haber, 1992), and it is likely that the human proteins have a similar function. It may be that the abolition of this second function is responsible for the more severe phenotype of the ERCC1 mouse, though this remains to be proven. It may also explain why only missense mutations have been found in XP-F patients. Null mutations may well not be compatible with life.

H. XP variant: A DNA polymerase involved in trans-lesion synthesis

In about 20% of XP patients, NER is normal, even though the patients exhibit classical XP symptoms (Robbins *et al.*, 1974) and their cells are UV-hypermutable just like classical XPs (Maher *et al.*, 1976; Myhr *et al.*, 1979). This group, designated XP variants, are not hypersensitive to killing by UV light (Arlett *et al.*, 1975). An early study found that XP-variant cells were defective in a process designated postreplication repair (Lehmann *et al.*, 1975). This is a mechanism or collection of mechanisms for generating high-molecular-weight daughter-strand DNA after UV irradiation, despite the presence of persisting UV damage in the parental DNA strands. Studies with *Escherichia coli* and lower eukaryotes suggest that there are likely to be different subpathways of postreplication repair. In yeast, for example, the major pathway is thought to involve recombination, whereas a minor pathway involves synthesis past the damage, a process designated trans-lesion synthesis (TLS) (Lawrence and Hinkle, 1996). The precise molecular defect in XP variants remained elusive for a long time. The lack of UV sensitivity precluded the cloning of the XP-variant gene by techniques involving UV selection. Recent work on *in vitro* systems has assisted in the elucidation of the underlying molecular mechanism. Work by several groups (Cordeiro-Stone *et al.*, 1997; Ensch-Simon *et al.*, 1998; Svoboda *et al.*, 1998) demonstrated that extracts of XP-variant cells were deficient in the ability to bypass a single thymine dimer in a DNA replication assay, thus mimicking the *in vivo* defect. By employing a system in which the template was a single-stranded plasmid containing a single chemical adduct, XP variants were shown to be unequivocally defective in TLS (Cordonnier *et al.*, 1999).

Recently the *XPV* gene has been cloned and found to be a homolog of the yeast *RAD30* gene (Masutani *et al.*, 1999a, 1999b; Johnson *et al.*, 1999a). The product of *S. cerevisiae RAD30* has been designated DNA polymerase η, and is involved in TLS (Johnson *et al.*, 1999b; McDonald *et al.*, 1997). *In vitro* studies show that both yeast and human proteins are able to insert adenines opposite a T-T CPD (Johnson *et al.*, 1999b; Masutani *et al.*, 1999b). Abolition of this relatively error-free pathway presumably channels photoproducts into more error-prone pathways, resulting in the increased mutability of XP-V cells. It is of interest that the UV mutation spectrum in XP variants is very different from that of normal and NER-defective XPs. Whereas the overriding mutations in the latter are C-to-T transitions, the XP variant spectrum contains similar numbers of transitions and transversions, at both G:C and A:T sites (Wang *et al.*, 1993).

III. COCKAYNE SYNDROME

A. Clinical and cellular characteristics

Clinically, CS exhibits characteristics quite different from most XP patients, although the neurological abnormalities are quite similar to the more extreme cases of XP. CS is characterized by dwarfism, loss of adipose tissue, mental retardation, gait defects, dental caries, pigmentary retinopathy, and acute sun sensitivity (Nance and Berry, 1992). Patients with CS also exhibit neurological defects, but the neurological changes in CS are demyelination whereas degeneration of the neurons prevails in XP. Furthermore, both CS and XP patients show ocular alterations. However, CS patients often develop cataracts and pigmentary retinopathy, features that are not associated with XP. Cells from patients with CS are, like XP cells, UV-hypersensitive. CS cells are, however, proficient in GGR, but they fail to recover RNA synthesis after UV irradiation (Mayne and Lehmann, 1982). This is associated with a specific defect in TCR (van Hoffen *et al.*, 1993). CS is therefore the counterpart of XP-C. The latter are able to carry out TCR but not GGR (see Section II.A). There are two CS complementation groups (CS-A and CS-B), the latter being the major group (Stefanini *et al.*, 1996), but as mentioned before, CS symptoms can, in rare cases, be associated with mutations in the *XPB*, *XPD*, and the *XPG* gene, where they coexist with XP symptoms. The defect in TCR indicates a possible role of the *CSA* and *CSB* genes in transcription, and it has been proposed that CS may be a disease that is caused by subtle defects in basal transcription (Friedberg, 1996; van Gool *et al.*, 1997b).

The *CSA* and *CSB* genes have been cloned (Henning *et al.*, 1995; Troelstra *et al.*, 1992). The *CSA* gene encodes a WD repeat protein, but there is as yet little information on its specific functions. The CSB protein has been found to contain helicase domains and is a member of the so-called Swi2/Snf2 DNA-dependent ATPase protein family (Eisen *et al.*, 1995). Despite the seven characteristic helicase domains found in all family members, helicase activity has not been demonstrated in any of this family of proteins, including CSB (Selby and Sancar, 1997b). They are currently thought to be involved in "chromatin remodeling" by disrupting protein–DNA interactions in chromatin (Pazin and Kadonaga, 1997).

B. CS and transcription

What is the role of the CS proteins in TCR? Initial hypotheses proposed that they were transcription-repair coupling factors, which recruited the NER machinery at sites where RNA polymerase molecules were blocked at lesions. In *E. coli* the product of the *mfd* gene performs such a role (Selby and Sancar, 1994), but direct evidence in support of a similar function for the CS proteins is lacking. The failure of CS cells to carry out TCR was proposed as an explanation for the characteristic inability of the cells to restore normal rates of RNA synthesis following UV irradiation (Mayne and Lehmann, 1982). However, if N-acetoxy-2-acetylaminofluorene rather than UV is used as the DNA-damaging agent, CS cells are sensitive to its lethal effects and RNA synthesis fails to recover, but there is no detectable defect in TCR. Van Oosterwijk *et al.* (1996) therefore proposed an alternative model in which, during NER, TFIIH is switched from its transcriptional to its NER role. After repair of damage, the CS proteins are required to switch TFIIH back from repair mode to transcription mode, thereby restoring RNA synthesis.

There are several reports of interactions between CS proteins and other repair proteins, for example, CSA with the p44 subunit of TFIIH and CSB with CSA, XPA, XPB, XPG, and the p34 subunit of TFIIE (Henning *et al.*, 1995; Iyer *et al.*, 1996). However, in size fractionation experiments with cell extracts, CSB and CSA ran as separate high-molecular-weight complexes, which did not contain any of the other repair proteins. However, about 10% of RNA polymerase II co-chromatographed and co-immunoprecipitated with CSB (van Gool *et al.*, 1997a). Several other reports have provided evidence for an interaction of CSB with RNA polymerase II. In one study CSB was found to increase the processivity of RNA polymerase II and added one extra nucleotide to a transcript blocked by a T-T CPD (Selby and Sancar, 1997a). CSB was also found to bind to ternary DNA-RNA-RNA polymerase II complexes (Tantin *et al.*, 1997), and the resulting quaternary complexes were able to recruit TFIIH (Tantin, 1998). These results suggest a possible role for the CSB protein in assisting RNA polymerase II at positions at which it is stalled.

In another study, UV irradiation was found to result in the ubiquitination of the large subunit of RNA polymerase II in normal cells but not in CS cells. The authors proposed that the function of the CSB proteins is to bring about the ubiquitin-mediated degradation of RNA polymerase II molecules stalled at lesions to enable repair to proceed (Bregman *et al.*, 1996). CS-B cells have also been reported to have reduced levels of transcription in both intact and perme-abilized cells (Balajee *et al.*, 1997). There is thus a considerable body of evidence implicating the CS proteins as intermediaries between repair and transcriptional processes, but their precise mode of action is still unclear.

A substantial amount of work (Bhatia *et al.*, 1996; Henning *et al.*, 1995; Troelstra *et al.*, 1992; van Gool *et al.*, 1994) indicates that neither CSA nor CSB is an essential gene. Many mutations in patients produce drastic truncations of the CS proteins (Mallery *et al.*, 1998). Furthermore, CSB knockout mice are viable and healthy. Although showing hypersensitivity to UV like CS patients, the mice have only mild neurological abnormalities, in contrast to the severe features of human CS-B individuals (van der Horst *et al.*, 1997). These observations clearly rule out an essential role of the CSB proteins in transcription, but a nonessential, accessory role for CS proteins has been proposed (van Gool *et al.*, 1997b). This would be in line with the dispensability of CS proteins for transcription, and the Swi2/Snf2-like domains could indicate that this may be carried out by disruption of an interaction between the stalled transcription machinery, possibly RNA polymerase II.

C. Oxidative damage

CS cells like those from XP-G/CS patients discussed in Section II.F are defective in transcription-coupled repair of thymine glycols (Leadon and Cooper, 1993). Work by Le Page *et al.* (2000) examined the mutagenic effect of a single oxidative lesion, 8-hydroxyguanine, in a plasmid molecule, when transfected into different cells. 8-Hydroxyguanine, if unrepaired, is able to base-pair with adenine, resulting in a G:C to A:T transversion. When transfected into normal or XP cells, only 4% of the lesions resulted in mutations, indicating that in most cases the 8-hydroxyguanine was efficiently repaired. In contrast, when transfected into either CS-B, XP-B/CS, XP-D/CS, or XP-G/CS cells, the mutation frequency was much higher, about 60% of the lesions generating mutations if the 8-hydroxyguanine was on the transcribed strand, whereas the basal level of 4% was obtained if the lesion was on the nontranscribed strand. These findings suggest that CS cells are defective in transcription-coupled repair of 8-hydroxyguanine, even though this lesion, when on the nontranscribed strand, is repaired efficiently by base excision repair. The relationship of these observations to the stimulation of hNth1 activity by XPG (see Section II.F) is unclear. XPG does not affect the activity of hOGG1, the protein responsible for the repair of 8-hydroxyguanine.

IV. TRICHOTHIODYSTROPHY

Trichothiodystrophy (TTD) is a heterogeneous disease characterized by abnormalities in the skin, teeth, nails, and the nervous system. Patients with TTD show growth abnormalities, mental retardation, and birdlike facies. The most prominent feature, which is also employed to secure the diagnosis, is brittle hair. This is caused by a reduction in the levels of cysteine-rich matrix proteins, leading to a "tiger-tail" appearance of the hair in polarized light microscopy (Itin and Pittelkow, 1990). Patients can be severely affected, leading to their death as early as 36 months, but there are also patients that have reached their 30s (Botta *et al.*, 1998). Many but not all patients exhibit sun sensitivity, and NER in cells varies widely [for reviews see Lehmann (1987, 1995); Stefanini *et al.* (1993)]. TTD patients have been assigned to three complementation groups. One is designated TTD-A, but the causative gene has not been isolated. The second is the XP-B group (one family), but the majority of cases (more than 25 reported in the literature) are in the XP-D complementation group. As described in Section II.E, evidence suggests that the site of the mutation in the *XPD* gene determines the phenotype (Taylor *et al.*, 1997), and these mutations are either found in the very N-terminal or the C-terminal region of the gene. We have suggested above that the transcriptional role of XPD is maintaining the structure of the TFIIH holo-complex, and that TTD features are largely the result of a transcriptional deficiency. It is likely that each mutation might affect the TFIIH structure to a different extent, and that mutations in which it is more severely disrupted will result in the most severe cases. An extra layer of complexity became apparent from a study of TTD patients in Italy (Botta *et al.*, 1998), which revealed a gene dosage effect. The mutation in the majority of these patients was R112H in the N-terminal region, and although all patients had similarly pronounced defects in NER, their clinical features were very heterogeneous. More detailed examination revealed that the more mildly affected cases were homozygous for this mutation, whereas the severely affected cases were functionally hemizygous, the second allele being null. This suggests that under conditions in which the XPD protein is partially crippled, its transcriptional role becomes rate-limiting, so that a twofold reduction in its level results in more severe effects.

If, as seems likely, the features of TTD are indeed the result of transcriptional deficiencies, how is the specificity of the features explained? De Boer *et al.* (1998) have proposed that the transcriptional deficiency becomes manifest in cells at the end of a differentiation pathway, when there is a high transcriptional load, but the transcription machinery becomes limiting. This would explain the specific deficiency in sulfur-rich matrix proteins of the hair shafts. In support of their ideas, they found that keratinocytes from the TTD mouse were deficient in transcription of the *SPRR2* gene, which encodes a proline-rich protein expressed late in terminal differentiation of interfollicular keratinocytes.

V. OUTSTANDING CLINICAL QUESTIONS

The lack of skin cancers in CS and TTD patients remains an enigma. In some cases of TTD, the repair deficiency is as marked as that in XP patients, and some patients have survived into their 30s with no sign of skin cancers, or indeed any of the other cutaneous abnormalities associated with XP. A contributing factor could be the attenuation of the UV dose by the thick skin associated with the ichthyosis found in TTD patients, but it seems unlikely that this could account for the complete lack of cutaneous pigmentation changes. The UV-induced mutation frequency is similarly elevated in XP-D and TTD fibroblasts, using either an endogenous gene

(Lehmann *et al.*, 1988) or a shuttle vector (Marionnet *et al.*, 1995) as mutation target. In the latter study, the mutation spectra for the XP and TTD cells used differed slightly, but the significance of this observation is not obvious.

It has been proposed that differences in the immune response may play a role in carcinogenesis in XP and TTD. XP-D patients and their cells can indeed show an impaired immune reaction (Gaspari *et al.*, 1993; Norris *et al.*, 1990) [see Bridges (1998) for review], but this is not found uniformly. A recent study by Ahrens *et al.* (1997) investigated the inhibition by UV irradiation of the expression of intercellular adhesion molecule (ICAM)-1, which is involved in immune responses. In this study, the inhibition of ICAM-1 expression in cells from three TTD patients was similar to that in normal cells, whereas in cells from three XP-D patients the inhibition was much greater. This suggested that the ICAM-1 expression correlated with the skin cancer risk. Extension of these findings to other patients, however, showed that although these findings pertained to the majority of XP and TTD cell strains, there were again some exceptions (Berneburg *et al.*, 2000a).

Another study examined the risk and the age of onset of basal cell carcinomas (BCC) in patients, who are frequently treated with UV phototherapy. Since this treatment is generally carried out for several months per year, it has been proposed that these patients are at an increased risk of developing skin cancer. The study by Dybdahl *et al.* (1999) looked at two polymorphisms in the *XPD* gene and found that one polymorphism conferred a higher risk of developing BCCs and that the age of onset of BCCs in individuals with this polymorphism was significantly lower than in other individuals. This study was limited in size and the findings must be regarded as preliminary, but it suggested that the *XPD* gene might play a causative role in the generation of skin cancers not only in patients with XP but also in the normal population.

A further series of studies measured the levels of catalase activities in XP and TTD cell strains. Primary XP cell strains from different complementation groups had very low levels of catalase activities, whereas TTD cell extracts had normal catalase (Vuillaume *et al.*, 1986, 1992). This appeared to be a post-transcriptional defect, since catalase mRNA levels were comparable in all cell strains, and the catalase defect could be rescued in XP-D cells transduced with *XPD* cDNA (Quilliet *et al.*, 1997). Other work has shown that the low catalase activity is caused by reduced intracellular levels of its cofactor NADPH in XP cells (Hoffschir *et al.*, 1998). The relationship of this defect to the other abnormalities in XP cells remains unexplained.

An altered apoptotic response to UV damage has also been proposed as a possible explanation for the differences in carcinogenic responses between XP-D and TTD patients. One hypothesis proposed that, in response to damage, XP cells might show reduced apoptosis relative to TTD, so that genetically unstable cells were not eliminated from the population in XP skin and were at greater risk for oncogenic transformation. A converse hypothesis proposes that excess apoptosis in XP versus TTD cells will lead to greater cell proliferation in the former to replace the apoptotic cells, and it is this greater proliferation that poses an increased carcinogenic risk [see Brash (1997) for discussion of the role of p53 mutations and apoptosis in skin cancer]. Apoptosis is mediated by p53 and p53 is upregulated by DNA damage. The p53 and apoptotic responses to UV irradiation of fibroblasts from XP and TTD patients appear to correlate with their repair capacity rather than their clinical features (Dumaz *et al.*, 1997; Ljungman and Zhang, 1996; Yamaizumi and Sugano, 1994). Keratinocytes from XP donors show increased apoptosis relative to normal keratinocytes after exposure to UV-B irradiation (C. Petit-Frere *et al.*, 2000), but there are as yet no published data on corresponding experiments with TTD patients, due to the difficulty of obtaining appropriate material.

The clinical features of XP and TTD appear to be exclusive. As described above, there are a few cases of joint symptoms of XP with CS in the XP-B, D, and G complementation groups. However, there are no reports in the literature of joint symptoms of XP and TTD. There seems to be some consequence of the mutations that result in TTD that prevents the manifestation of XP skin abnormalities, despite very low levels of NER. Since the clinical features of TTD seem to be unrelated to the repair levels, and the latter can range from close to zero up to normal, it seems unlikely that the lack of XP features in TTD has anything to do with differing levels of repair between XP and TTD patients (Berneburg *et al.*, 2000a). It is more likely that the subtle transcriptional deficiencies in TTD patients in some way reduce the potential of unrepaired DNA damage to act as a precarcinogenic lesion. In contrast to the human situation, both the TTD mouse and the CS-B knockout mice are hypersensitive to UVB-induced carcinogenesis (de Boer *et al.*, 1999; van der Horst *et al.*, 1997), albeit to a lesser extent than the XP-A mice. Clearly, more research is required to unravel the relationships among these complex phenomena.

Another clinical attribute that awaits a satisfactory explanation is the neurological abnormalities associated in XP most severely with XP-A patients and to a lesser extent with XP-D and XP-G, but rarely if at all seen in XP-C, E, F, or variant patients. An early hypothesis suggested that, during life, the CNS is exposed to DNA-damaging agents, which generate damage that is normally repaired by NER (Robbins *et al.*, 1983). In the most severely deficient patients (XP-A) this damage is not repaired, resulting in neuronal death and the early onset of neurological abnormalities. In cells with some residual activity, like most XP-Ds, this only becomes manifest later. More recent models have suggested a role for transcriptional defects in the generation of neurological abnormalities and/or a role for oxidative damage. None of these models is entirely satisfactory, and unfortunately the XP knockout mice do not show neurological abnormalities and are therefore of limited use in answering this specific question.

VI. CONCLUDING REMARKS

Developments over the last few years have provided remarkable insights into the mechanism of nucleotide excision repair. They have also revealed unanticipated complexities. The discoveries of multifunctional roles of NER proteins in transcription (XPB and XPD), recombination (XPF), and repair of oxidative damage (XPG) provide novel insights into complicated functional interrelationships between these processes. The precise ways in which these pathways interact within the context of chromatin in the cell nucleus will be one of the challenges of the next few years. In contrast to the detailed understanding of global genome repair, we as yet understand very little of the specific mechanism of transcription-coupled repair and the functions of the CS proteins. The ultimate goal of researchers in this area will be to understand the basis of the very marked differences in the clinical features of three disorders with apparently similar molecular defects.

References

Abramic, M., Levine, A. S., and Protic, M. (1991). Purification of an ultraviolet-inducible, damage-specific DNA-binding protein from primate cells. *J. Biol. Chem.* **266,** 22493–22500.

Ahrens, C., Grewe, M., Berneburg, M., Grether-Beck, S., Quilliet, S., Mezzina, M., Sarasin, A., Lehmann, A. R., Arlett, C. F., and Krutmann, J. (1997). Photocarcinogenesis correlates with immunosuppression in DNA-repair-defective individuals. *Proc. Natl. Acad. Sci. (USA)* **94,** 6837–6841.

Arlett, C. F., Harcourt, S. A., and Broughton, B. C. (1975). The influence of caffeine on cell survival in excision-proficient and excision-deficient xeroderma pigmentosum and normal human cell strains following ultraviolet light irradiation. *Mutation Res.* **33,** 341–346.

Asahina, H., Kuraoka, I., Shirakawa, M., Morita, E. H., Miura, N., Miyamoto, I., Ohtsuka, E., Okada, Y., and Tanaka, K. (1994). The XPA protein is a zinc metalloprotein with an ability to recognize various kinds of DNA damage. *Mutation Res.* **315,** 229–237.

Balajee, A. S., May, A., Dianov, G. L., Friedberg, E. C., and Bohr, V. A. (1997). Reduced RNA polymerase II transcription in intact and permeabilized Cockayne syndrome group B cells. *Proc. Natl. Acad. Sci. (USA)* **94,** 4306–4311.

Baxter, B. K., and Smerdon, M. J. (1998). Nucleosome unfolding during DNA repair in normal and xeroderma pigmentosum (group C) human cells. *J. Biol. Chem.* **273,** 17517–17524.

Berg, R. J., Ruven, H. J., Sands, A. T., de Gruijl, F. R., and Mullenders, L. H. (1998). Defective global genome repair in XPC mice is associated with skin cancer susceptibility but not with sensitivity to UVB induced erythema and edema. *J. Invest. Dermatol.* **110,** 405–409.

Berneburg, M., Clingen, P. H., Harcourt, S. A., Lowe, J. E., Taylor, E. M., Green, M. H. L., Krutmann, J., Arlett, C. F., and Lehmann, A. R. (2000a). The cancer-free phenotype in trichothiodystrophy is unrelated to its repair defect. *Cancer Res.,* **60,** 431–438.

Berneburg, M., Lowe, J. E., Nardo, T., Araujo, S., Fousteri, M., Green, M. H. L., Krutmann, J., Wood, R. D., Stefanini, M., and Lehmann, A. R. (2000b). DNA damage causes uncontrolled DNA breakage in cells from patients with combined features of XP-D and Cockayne syndrome. *EMBO J.,* **19,** 1157–1162.

Bessho, T. (1999). Nucleotide excision repair 3′ endonuclease XPG stimulates the activity of base excision repairenzyme thymine glycol DNA glycosylase. *Nucleic Acids Res.* **27,** 979–983.

Bhatia, P. K., Verhage, R. A., Brouwer, J., and Friedberg, E. C. (1996). Molecular cloning and characterization of *Saccharomyces cerevisiae RAD28*, the yeast homolog of the human Cockayne syndrome A *(CSA)* gene. *J. Bacteriol.* **178,** 5977–5988.

Bootsma, D., and Hoeijmakers, J. H. J. (1993). Engagement with transcription. *Nature* **363,** 114–115.

Bootsma, D., Kraemer, K. H., Cleaver, J. E., and Hoeijmakers, J. H. J. (1998). Nucleotide excision repair syndromes: Xeroderma pigmentosum, Cockayne syndrome, and trichothiodystrophy. In "The Genetic Basis of Human Cancer" (B. Vogelstein, and K. W. Kinzler, eds.), pp. 245–274. McGraw-Hill, New York.

Botta, E., Nardo, T., Broughton, B. C., Marinoni, S., Lehmann, A. R., and Stefanini, M. (1998). Analysis of mutations in the *XPD* gene in Italian patients with trichothiodystrophy: Site of mutation correlates with repair deficiency but gene dosage appears to determine clinical severity. *Am. J. Hum. Genet.* **63,** 190–196.

Brash, D. E. (1997). Sunlight and the onset of skin cancer. *Trends Genet.* **13,** 410–414.

Bregman, D. B., Halaban, R., van Gool, A. J., Henning, K. A., Friedberg, E. C., and Warren, S. L. (1996). UV-induced ubiquitination of RNA polymerase II: A novel modification deficient in Cockayne syndrome cells. *Proc. Natl. Acad. Sci. (USA)* **93,** 11586–11590.

Bridges, B. A. (1998). UV-induced mutations and skin cancer: How important is the link? *Mutation Res.* **422,** 23–30.

Buchko, G. W., Iakoucheva, L. M., Kennedy, M. A., Ackerman, E. J., and Hess, N. J. (1999). Extended X-ray absorption fine structure evidence for a single metal binding domain in *Xenopus laevis* nucleotide excision repair protein XPA. *Biochem. Biophys. Res. Commun.* **254,** 109–113.

Chavanne, F., Broughton, B. C., Pietra, D., Nardo, T., Browitt, A., Lehmann, A. R., and Stefanini, M. (2000). Mutations in the *XPC* gene in families with xeroderma pigmentosum and consequences at the cell, protein and transcript levels. *Cancer Res.* **60,** 1974–1982.

Cheo, D. L., Meira, L. B., Hammer, R. E., Burns, D. K., Doughty, A. T., and Friedberg, E. C. (1996). Synergistic interactions between XPC and p53 mutations in double-mutant mice: Neural tube abnormalities and accelerated UV radiation-induced skin cancer. *Curr. Biol.* **6,** 1691–1694.

Cheo, D. L., Ruven, H. J., Meira, L. B., Hammer, R. E., Burns, D. K., Tappe, N. J., van Zeeland, A. A., Mullenders, L. H., and Friedberg, E. C. (1997). Characterization of defective nucleotide excision repair in XPC mutant mice. *Mutat. Res.* **374,** 1–9.

Chu, G., and Chang, E., 1988, Xeroderma pigmentosum group E cells lack a nuclear factor that binds to damaged DNA. *Science* **242,** 564–567.

Cleaver, J. E., Thompson, L. H., Richardson, A. S., and States, J. C. (1999). A summary of mutations in the UV-sensitive disorders: Xeroderma pigmentosum, Cockayne syndrome, and trichothiodystrophy. *Hum. Mutat.* **14,** 9–22.

Cloud, K. G., Shen, B., Strniste, G. F., and Park, M. S. (1995). XPG protein has a structure-specific endonuclease activity. *Mutat. Res.* **347,** 55–60.

Coin, F., Bergmann, E., Tremeau-Bravard, A., and Egly, J. M. (1999). Mutations in XPB and XPD helicases found in xeroderma pigmentosum patients impair the transcription function of TFIIH. *EMBO. J.* **18,** 1357–1366.

Cooper, P. K., Nouspikel, T., Clarkson, S. G., and Leadon, S. A. (1997). Defective transcription-coupled repair of oxidative base damage in Cockayne syndrome patients from XP group G. *Science* **275,** 990–993.

Cordeiro-Stone, M., Zaritskaya, L. S., Price, L. K., and Kaufmann, W. K. (1997). Replication fork bypass of a pyrimidine dimer blocking leading strand DNA synthesis. *J. Biol. Chem.* **272,** 13945–13954.

Cordonnier, A. M., Lehmann, A. R., and Fuchs, R. P. P. (1999). Impaired translesion synthesis in xeroderma pigmentosum extracts. *Mol. Cell. Biol.* **19,** 2206–2211.

de Boer, J., de Wit, J., van Steeg, H., Berg, R. J. W., Morreau, H., Visser, P., Lehmann, A. R., Duran, M., Hoeijmakers, J. H. J., and Weeda, G. (1998). A mouse model for the basal transcription/DNA repair syndrome trichothiodystrophy. *Mol. Cell* **1,** 981–990.

de Boer, J., van Steeg, H., Berg, R. J., Garssen, J., de Wit, J., van Oostrum, C. T., Beems, R. B., van der Horst, G. T., van Kreijl, C. F., de Gruijl, F. R., Bootsma, D., Hoeijmakers, J. H., and Weeda, G. (1999). Mouse model for the DNA repair/basal transcription disorder trichothiodystrophy reveals cancer predisposition. *Cancer Res.* **59,** 3489–3494.

de Laat, W. L., Jaspers, N. G., and Hoeijmakers, J. H. (1999). Molecular mechanism of nucleotide excision repair. *Genes Dev.* **13,** 768–785.

de Vries, A., van Oostrom, C. T., Hofhuis, F. M., Dortant, P. M., Berg, R. J., de Gruijl, F. R., Wester, P. W., van Kreijl, C. F., Capel, P. J., van Steeg, H., and Verbeek, S. J. (1995). Increased susceptibility to ultraviolet-B and carcinogens of mice lacking the DNA excision repair gene XPA. *Nature* **377,** 169–173.

Drapkin, R., Reardon, J. T., Ansari, A., Huang, J. C., Zawel, L., Ahn, K., Sancar, A., and Reinberg, D. (1994). Dual role of TFIIH in DNA excision repair and in transcription by RNA polymerase II. *Nature* **368,** 769–772.

Drapkin, R., Le Roy, G., Cho, H., Akoulitchev, S., and Reinberg, D. (1996). Human cyclin-dependent kinase-activating kinase exists in three distinct complexes. *Proc. Natl. Acad. Sci. (USA)* **93,** 6488–6493.

Dumaz, N., Drougar, C., Sarasin, A., and Daya-Grosjean, L. (1993). Specific UV-induced mutation spectrum in the p53 gene of skin tumors from DNA repair deficient xeroderma pigmentosum patients. *Proc. Natl. Acad. Sci. (USA)* **90,** 10529–10533.

Dumaz, N., Duthu, A., Ehrhart, J. C., Drougard, C., Appella, E., Anderson, C. W., May, P., Sarasin, A., and Daya-Grosjean, L. (1997). Prolonged p53 protein accumulation in trichothiodystrophy fibroblasts dependent on unrepaired pyrimidine dimers on the transcribed strands of cellular genes. *Mol. Carcinog.* **20,** 340–347.

Dybdahl, M., Vogel, U., Frentz, G., Wallin, H., and Nexo, B. A. (1999). Polymorphisms in the DNA repair gene XPD: Correlations with risk and age at onset of basal cell carcinoma. *Cancer Epidemiol. Biomarkers Prev.* **8,** 77–81.

Eisen, J. A., Sweder, K. S., and Hanawalt, P. C. (1995). Evolution of the SNF2 family of proteins—Subfamilies with distinct sequences and functions. *Nucleic Acids Res.* **23,** 2715–2723.

Ensch-Simon, I., Burgers, P. M., and Taylor, J. S. (1998). Bypass of a site-specific cis-Syn thymine dimer in an SV40 vector during *in vitro* replication by HeLa and XPV cell-free extracts. *Biochemistry* **37,** 8218–8226.

Evans, E., Fellows, J., Coffer, A., and Wood, R. D. (1997a). Open complex formation around a lesion during nucleotide excision repair provides a structure for cleavage by human XPG protein. *EMBO J.* **16,** 625–638.

Evans, E., Moggs, J. G., Hwang, J. R., Egly, J. M., and Wood, R. D. (1997b). Mechanism of open complex and dual incision formation by human nucleotide excision repair factors. *EMBO J.* **16,** 6559–6573.

Feaver, W. J., Svejstrup, J. Q., Bardwell, L., Bardwell, A. J., Buratowski, S., Gulyas, K. D., Donahue, T. F., Friedberg, E. C., and Kornberg, R. D. (1993). Dual roles of a multiprotein complex from *S. cevisiae* in transcription and DNA repair. *Cell* **75,** 1379–1387.

Fishman-Lobell, J., and Haber, J. E. (1992). Removal of nonhomologous DNA ends in double-strand break recombination: The role of the yeast ultraviolet repair gene RAD1. *Science* **258,** 480–484.

Friedberg, E. C. (1996). Cockayne syndrome—a primary defect in DNA repair, transcription, both or neither? *BioEssays* **18,** 731–738.

Friedberg, E. C., Walker, G. C., and Siede, W. (1995). "DNA Repair and Mutagenesis." ASM Press, Washington, DC.

Gaspari, A. A., Fleisher, T. A., and Kraemer, K. H. (1993). Impaired interferon production and natural killer cell activation in patients with the skin cancer prone disorder, xeroderma pigmentosum. *J. Clin. Invest.* **92**, 1135–1142.

Giglia, G., Dumaz, N., Drougard, C., Avril, M. F., Daya-Grosjean, L., and Sarasin, A. (1998). p53 Mutations in skin and internal tumors of xeroderma pigmentosum patients belonging to the complementation group C. *Cancer Res.* **58**, 4402–4409.

Guzder, S. N., Sung, P., Bailly, V., Prakash, L., and Prakash, S. (1994). Rad25 is a DNA helicase required for RNA repair and RNA-polymerase-II transcription. *Nature* **369**, 578–581.

Harada, Y. N., Shiomi, N., Koike, M., Ikawa, M., Okabe, M., Hirota, S., Kitamura, Y., Kitagawa, M., Matsunaga, T., Nikaido, O., and Shiomi, T. (1999). Postnatal growth failure, short life span, and early onset of cellular senescence and subsequent immortalization in mice lacking the xeroderma pigmentosum group G gene. *Mol. Cell. Biol.* **19**, 2366–2372.

Henning, K. A., Li, L., Iyer, N., McDaniel, L. D., Reagan, M. S., Legerski, R., Schultz, R. A., Stefanini, M., Lehmann, A. R., Mayne, L. V., and Friedberg, E. C. (1995). The Cockayne-syndrome group-A gene encodes a WD repeat protein that interacts with CSB protein and a subunit of RNA-polymerase-II TFIIH. *Cell* **82**, 555–564.

Hess, N. J., Buchko, G. W., Conradson, S. D., Espinosa, F. J., Ni, S., Thrall, B. D., and Kennedy, M. A. (1998). Human nucleotide excision repair protein XPA: Extended X-ray absorption fine-structure evidence for a metal-binding domain. *Protein Sci.* **7**, 1970–1975.

Hoeijmakers, J. H. J., Egly, J.-M., and Vermeulen, W. (1996). TFIIH: A key component in multiple DNA transactions. *Curr. Opin. Gen. Dev.* **6**, 26–33.

Hoffschir, F., Daya-Grosjean, L., Petit, P. X., Nocentini, S., Dutrillaux, B., Sarasin, A., and Vuillaume, M. (1998). Low catalase activity in xeroderma pigmentosum fibroblasts and SV40-transformed human cell lines is directly related to decreased intracellular levels of the cofactor, NADPH. *Free Radic. Biol. Med.* **24**, 809–816.

Hwang, B. J., Toering, S., Francke, U., and Chu, G. (1998). p48 Activates a UV-damaged DNA-binding factor and is defective in xeroderma pigmentosum group E cells that lack binding activity. *Mol. Cell. Biol.* **18**, 4391–4399

Hwang, B. J., Ford, J. M., Hanawalt, P. C., and Chu, G. (1999). Expression of the p48 xeroderma pigmentosum gene is p53-dependent and is involved in global genomic repair. *Proc. Natl. Acad. Sci. (USA)* **96**, 424–428.

Itin, P. H., and Pittelkow, M. R., 1990, Trichothiodystrophy: Review of sulfur-deficient brittle hair syndromes and association with the ectodermal dysplasias, *J. Am. Acad. Dermatol.* **20**, 705–717.

Iyer, N., Reagan, M. S., Wu, K. J., Canagarajah, B., and Friedberg, E. C. (1996). Interactions involving the human RNA polymerase II transcription/nucleotide excision repair complex TFIIH, the nucleotide excision repair protein XPG, and Cockayne syndrome group B (CSB) protein. *Biochemistry* **35**, 2157–2167.

Johnson, R. E., Kondratick, C. M., Prakash, S., and Prakash, L. (1999a). hRAD30 Mutations in the variant form of xeroderma pigmentosum. *Science* **285**, 263–265.

Johnson, R. E., Prakash, S., and Prakash, L. (1999b). Efficient bypass of a thymine-thymine dimer by yeast DNA polymerase, Polη. *Science* **283**, 1001–1004.

Jones, C. J., and Wood, R. D. (1993). Preferential binding of the xeroderma pigmentosum group A complementing protein to damaged DNA. *Biochemistry* **32**, 12096–12104.

Keeney, S., Wein, H., and Linn, S. (1992). Biochemical heterogeneity in xeroderma pigmentosum complementation group E. *Mutat. Res.* **273**, 49–56.

Keeney, S., Chang, G. J., and Linn, S. (1993). Characterization of a human DNA damage binding protein implicated in xeroderma pigmentosum E. *J. Biol. Chem.* **268**, 21293–21300.

Klungland, A., Hoss, M., Gunz, D., Constantinou, A., Clarkson, S. G., Doetsch, P. W., Bolton, P. H., Wood, R. D., and Lindahl, T. (1999). Base excision repair of oxidative DNA damage activated by XPG protein. *Mol. Cell* **3**, 33–42.

Kraemer, K. H., Lee, M. M., and Scotto, K. (1984). DNA repair protects against cutaneous and internal neoplasia: Evidence from xeroderma pigmentosum. *Carcinogenesis* **5,** 511–514.

Kraemer, K. H., Lee, M. M., and Scotto, J. (1987). Xeroderma pigmentosum. Cutaneous, ocular and neurologic abnormalities in 830 published cases. *Arch. Dermatol.* **123,** 241–250.

Lawrence, C. W., and Hinkle, D. C. (1996). DNA polymerase zeta and the control of DNA damage induced mutagenesis in eukaryotes. *Cancer Surv.* **28,** 21–31.

Le Page, F., Kwoh, E. E., Avrutskaya, A., Gentil, A., Leadon, S. A., Sarasin, A. and Cooper, P. K. (2000). Transcription-coupled repair of 8-oxoguanine: requirement for XPG, TFIIH, and CSB and implications for Cockayne syndrome. *Cell* **101,** 159–171.

Leadon, S. A., and Cooper, P. K. (1993). Preferential repair of ionizing radiation-induced damage in the transcribed strand of an active human gene is defective in Cockayne syndrome. *Proc. Natl. Acad. Sci. (USA)* **90,** 10499–10503.

Lehmann, A. R. (1987). Cockayne's syndrome and trichothiodystrophy: Defective repair without cancer. *Cancer Rev.* **7,** 82–103.

Lehmann, A. R. (1995). Nucleotide excision repair and the link with transcription. *Trends Biochem. Sci.* **20,** 402–405.

Lehmann, A. R. (1998). Dual functions of DNA repair genes: Molecular, cellular, and clinical implications. *BioEssays* **20,** 146–155.

Lehmann, A. R., Kirk-Bell, S., Arlett, C. F., Paterson, M. C., Lohman, P. H. M., de Weerd-Kastelein, E. A., and Bootsma, D. (1975). Xeroderma pigmentosum cells with normal levels of excision repair have a defect in DNA synthesis after UV-irradiation. *Proc. Natl. Acad. Sci. (USA)* **72,** 219–223.

Lehmann, A. R., Arlett, C. F., Broughton, B. C., Harcourt, S. A., Steingrimsdottir, H., Stefanini, M., Taylor, A. M. R., Natarajan, A. T., Green, S., King, M. D., MacKie, R. M., Stephenson, J. B. P., and Tolmie, J. L. (1988). Trichothiodystrophy, a human DNA repair disorder with heterogeneity in the cellular response to ultraviolet light. *Cancer Res.* **48,** 6090–6096.

Li, L., Bales, E. S., Peterson, C. A., and Legerski, R. J. (1993). Characterization of molecular defects in xeroderma pigmentosum group C. *Nature Genet.* **5,** 413–417.

Li, R. Y., Calsou, P., Jones, C. J., and Salles, B. (1998). Interactions of the transcription/DNA repair factor TFIIH and XP repair proteins with DNA lesions in a cell-free repair assay. *J. Mol. Biol.* **281,** 211–218.

Ljungman, M., and Zhang, F. (1996). Blockage of RNA polymerase as a possible trigger for u.v. light-induced apoptosis. *Oncogene* **13,** 823–831.

Ma, L., Siemssen, E. D., Noteborn, M., and Van der Eb, A. J. (1994). The xeroderma pigmentosum group B protein ERCC3 produced in the baculovirus system exhibits DNA helicase activity. *Nucleic. Acids Res.* **22,** 4095–4102.

Maher, V. M., Ouellette, L. M., Curren, R. D., and McCormick, J. J. (1976). Frequency of ultraviolet light-induced mutations is higher in xeroderma pigmentosum variant cells than in normal human cells. *Nature* **261,** 593–595.

Mallery, D. L., Tanganelli, B., Colella, S., Steingrimsdottir, H., Van Gool, A. J., Troelstra, C., Stefanini, M., and Lehmann, A. R. (1998). Molecular analysis of mutations in the *CSB(ERCC6)* gene in patients with Cockayne syndrome. *Am. J. Hum. Genet.* **62,** 77–85.

Marionnet, C., Benoit, A., Benhamou, S., Sarasin, A., and Stary, A. (1995). Characteristics of UV-induced mutation spectra in human XP-D/ERCC2 gene-mutated xeroderma pigmentosum and trichothiodystrophy cells. *J. Mol. Biol.* **252,** 550–562.

Masutani, C., Sugasawa, K., Yanagisawa, J., Sonoyama, T., Ui, M., Enomoto, T., Takio, K., Tanaka, K., van der Spek, P. J., Bootsma, D., Hoeijmakers, J. H. J., and Hanaoka, F. (1994). Purification and cloning of a nucleotide excision-repair complex involving the xeroderma-pigmentosum group-C protein and a human homolog of yeast RAD23. *EMBO J.* **13,** 1831–1843.

Masutani, C., Araki, M., Yamada, A., Kusumoto, R., Nogimori, T., Maekawa, T., Iwai, S., and Hanaoka, F. (1999a). Xeroderma pigmentosum variant (XP-V) correcting protein from HeLa cells has a thymine dimer bypass DNA polymerase activity. *EMBO J.* **18,** 3491–3501.

Masutani, C., Kusumoto, R., Yamada, A., Dohmae, N., Yokoi, M., Yuasa, M., Araki, M., Iwai, S., Takio, K., and Hanaoka, F. (1999b). The human XPV (Xeroderma Pigmentosum Variant) gene encodes human polymerase η. *Nature* **399**, 700–704.

Matsumura, Y., Sato, M., Nishigori, C., Zghal, M., Yagi, T., Imamura, S., and Takebe, H. (1995). High prevalence of mutations in the p53 gene in poorly differentiated squamous cell carcinomas in xeroderma pigmentosum patients. *J. Invest. Dermatol.* **105**, 399–401.

Mayne, L. V., and Lehmann, A. R. (1982). Failure of RNA synthesis to recover after UV-irradiation: An early defect in cells from individuals with Cockayne's syndrome and xeroderma pigmentosum. *Cancer Res.* **42**, 1473–1478.

McDonald, J. P., Levine, A. S., and Woodgate, R. (1997). The *Saccharomyces cerevisiae RAD30* gene, a homologue of *Escherichia coli dinB* and *umuC*, is DNA damage inducible and functions in a novel error-free postreplication repair mechanism. *Genetics* **147**, 1557–1568.

McWhir, J., Selfridge, J., Harrison, D. J., Squires, S., and Melton, D. W. (1993). Mice with DNA repair gene (*ERCC-1*) deficiency have elevated levels of p53, liver nuclear abnormalities and die before weaning. *Nature Genet.* **5**, 217–223.

Morita, E. H., Ohkubo, T., Kuraoka, I., Shirakawa, M., Tanaka, K., and Morikawa, K. (1996). Implications of the zinc-finger motif found in the DNA-binding domain of the human XPA protein. *Genes Cells* **1**, 437–442.

Mu, D., and Sancar, A. (1997). Model for XPC-independent transcription-coupled repair of pyrimidine dimers in humans. *J. Biol. Chem.* **272**, 7570–7573.

Myhr, B. C., Turnbull, D., and di Paulo, J. A. (1979). Ultraviolet mutagenesis of normal and xeroderma pigmentosum variant human fibroblasts. *Mutat. Res.* **62**, 341–353.

Nakane, H., Takeuchi, S., Yuba, S., Sijo, M., Nakatsu, Y., Murai, H., Nakatsuru, Y., Ishikawa, T., Hirota, S., Kitamura, Y., Kato, Y., Tsunoda, Y., Miyauchi, H., Horio, T., Tokunaga, T., Matsunaga, T., Nikaido, O., Nishimune, Y., Okada, Y., and Tanaka, K. (1995). High incidence of ultraviolet-B- or chemical-carcinogen-induced skin tumours in mice lacking the xeroderma pigmentosum group A gene. *Nature* **377**, 165–168.

Nance, M. A., and Berry, S. A. (1992). Cockayne syndrome: Review of 140 cases. *Am. J. Med. Genet.* **42**, 68–84.

Nichols, A. F., Ong, P., and Linn, S. (1996). Mutations specific to the xeroderma pigmentosum group E Ddb-phenotype. *J. Biol. Chem.* **40**, 24317–24320.

Nishigori, C., Zghal, M., Yagi, T., Imamura, S., Komoun, M. R., and Takebe, H. (1993). High prevalence of the point mutation in exon 6 of the xeroderma pigmentosum group A-complementing (XPAC) gene in xeroderma pigmentosum group A patients in Tunisia. *Am. J. Hum. Genet.* **53**, 1001–1006.

Nishigori, C., Moriwaki, S.-I., Takebe, H., Tanaka, T., and Imamura, S. (1994). Gene alterations and clinical characteristics of xeroderma pigmentosum group A patients in Japan. *Arch. Dermatol.* **130**, 191–197.

Nocentini, S., Coin, F., Saijo, M., Tanaka, K., and Egly, J. M. (1997). DNA damage recognition by XPA protein promotes efficient recruitment of transcription factor II H. *J. Biol. Chem.* **272**, 22991–22994.

Norris, P. G., Limb, G. A., Hamblin, A. S., Lehmann, A. R., Arlett, C. F., Cole, J., Waugh, A. P. W., and Hawk, J. L. M. (1990). Immune function, mutant frequency and cancer risk in the DNA repair defective genodermatoses xeroderma pigmentosum, Cockayne's syndrome and trichothiodystrophy. *J. Invest. Dermatol.* **94**, 94–100.

Nouspikel, T., Lalle, P., Leadon, S. A., Cooper, P. K., and Clarkson, S. G. (1997). A common mutational pattern in Cockayne syndrome patients from xeroderma pigmentosum group G: Implications for a second XPG function. *Proc Natl Acad Sci (USA)* **94**, 3116–3121.

O'Donovan, A., Davies, A. A., Moggs, J. G., West, S. C., and Wood, R. D. (1994). XPG endonuclease makes the 3′ incision in human DNA nucleotide excision repair. *Nature* **371**, 432–435.

Park, E., Guzder, S. N., Koken, M. H. M., Jaspers-Dekker, I., Weeda, G., Hoeijmakers, J. H. J., Prakash, S., and Prakash, L. (1992). RAD25 (SSL2), the yeast homolog of the human xeroderma pigmentosum group B DNA repair gene, is essential for viability. *Proc. Natl. Acad. Sci. (USA)* **89**, 11416–11420.

Pazin, M. J., and Kadonaga, J. T. (1997). SWI2/SNF2 and related proteins: ATP-driven motors that disrupt protein-DNA interactions. *Cell* **88**, 737–740.

Petit-Frere, C., Capulas, E., Lowe, J. E., Koulu, L., Marttila, R. J., Jaspers, N. G. J., Clingen, P. H., Green, M. H. L. and Arlett, C. F. (2000). Ultraviolet-B-induced apoptosis and cytokine release in xeroderma pigmentosum keratinocytes. *J. Invest. Dermatol.*, in press.

Quilliet, X., Chevallier-Lagente, O., Zeng, L., Calvayrac, R., Mezzina, M., Sarasin, A., and Vuillaume, M. (1997). Retroviral-mediated correction of DNA repair defect in xeroderma pigmentosum cells is associated with recovery of catalase activity. *Mutat. Res.* **385**, 235–242.

Rapic Otrin, V., Kuraoka, I., Nardo, T., McLenigan, M., Eker, A. P., Stefanini, M., Levine, A. S., and Wood, R. D. (1998). Relationship of the xeroderma pigmentosum group E DNA repair defect to the chromatin and DNA binding proteins UV-DDB and replication protein A. *Mol. Cell. Biol.* **18**, 3182–3190.

Reardon, J. T., Ge, H., Gibbs, E., Sancar, A., Hurwitz, J., and Pan, Z. Q. (1996). Isolation and characterization of two human transcription factor IIH (TFIIH)-related complexes: ERCC2/CAK and TFIIH [published erratum appears in *Proc. Natl. Acad. Sci. (USA)* **93**, 10538 (1996)]. *Proc. Natl. Acad. Sci. (USA)* **93**, 6482–6487.

Robbins, J. H., Kraemer, K. H., Lutzner, M. A., Festoff, B. W., and Coon, H. G. (1974). Xeroderma pigmentosum: An inherited disease with sun-sensitivity, multiple cutaneous neoplasms, and abnormal DNA repair. *Ann. Intern. Med.* **80**, 221–248.

Robbins, J. H., Polinsky, R. J., and Moshell, A. N. (1983). Evidence that lack of deoxyribonucleic acid repair causes death of neurons in xeroderma pigmentosum. *Ann. Neurol.* **13**, 682–684.

Sands, A. T., Abuin, A., Sanchez, A., Conti, C. J., and Bradley, A. (1995). High susceptibility to ultraviolet-induced carcinogenesis in mice lacking XPC. *Nature* **377**, 162–165.

Schaeffer, L., Roy, R., Humbert, S., Moncollin, V., Vermeulen, W., Hoeijmakers, J. H. J., Chambon, P., and Egly, J.-M. (1993). DNA repair helicase: A component of BTF2 (TFIIH) basic transcription factor. *Science* **260**, 58–63.

Schaeffer, L., Monocollin, V., Roy, R., Staub, A., Mezzina, M., Sarasin, A., Weeda, G., Hoeijmakers, J. H. J., and Egly, J. M. (1994). The ERCC2/DNA repair protein is associated with the class II BTF2/TFIIH transcription factor. *EMBO J.* **13**, 2388–2392.

Selby, C. P., and Sancar, A. (1994). Mechanisms of transcription-repair coupling and mutation frequency decline. *Microbiol. Rev.* **58**, 317–329.

Selby, C. P., and Sancar, A. (1997a). Cockayne syndrome group B protein enhances elongation by RNA polymerase II. *Proc. Natl. Acad. Sci. (USA)* **95**, 11205–11209.

Selby, C. P., and Sancar, A. (1997b). Human transcription-repair coupling factor CSB/ERCC6 is a DNA-stimulated ATPase but is not a helicase and does not disrupt the ternary transcription complex of stalled RNA polymerase II. *J. Biol. Chem.* **272**, 1885–1890.

Sijbers, A. M., de Laat, W. L., Ariza, R. R., Biggerstaff, M., Wei, Y.-F., Moggs, J. G., Carter, K. C., Shell, B. K., Evans, E., de Jong, M. C., Rademakers, S., de Rooij, J., Jaspers, N. G. J., Hoeijmakers, J. H. J., and Wood, R. D. (1996). Xerderma pigmentosum group F caused by a defect in a structure-specific DNA repair endonuclease. *Cell* **86**, 811–822.

States, J. C., McDuffie, E. R., Myrand, S. P., McDowell, M., and Cleaver, J. E. (1998). Distribution of mutations in the human xeroderma pigmentosum group A gene and their relationships to the functional regions of the DNA damage recognition protein. *Hum. Mutat.* **12**, 103–113.

Stefanini, M., Lagomarsini, P., Giliani, S., Nardo, T., Botta, E., Peserico, A., Kleijer, W. J., Lehmann, A. R., and Sarasin, A. (1993). Genetic heterogeneity of the excision repair defect associated with trichothiodystrophy. *Carcinogenesis* **14**, 1101–1105.

Stefanini, M., Fawcett, H., Botta, E., Nardo, T., and Lehmann, A. R. (1996). Genetic analysis of twenty-two patients with Cockayne syndrome. *Hum. Genet.* **97,** 418–423.

Sugasawa, K., Ng, J. M., Masutani, C., Iwai, S., van der Spek, P. J., Eker, A. P., Hanaoka, F., Bootsma, D., and Hoeijmakers, J. H. (1998). Xeroderma pigmentosum group C protein complex is the initiator of global genome nucleotide excision repair. *Mol. Cell* **2,** 223–232.

Sung, P., Higgins, D., Prakash, L., and Prakash, S. (1988). Mutation of lysine-48 to arginine in the yeast RAD3 protein abolishes its ATPase and DNA helicase activities but not the ability to bind ATP. *EMBO J.* **7,** 3263–3269.

Svoboda, D. L., Briley, L. P., and Vos, J. M. (1998). Defective bypass replication of a leading strand cyclobutane thymine dimer in xeroderma pigmentosum variant cell extracts. *Cancer Res.* **58,** 2445–2448.

Takebe, H., Miki, Y., Kozuka, T., Furuyama, J. I., Tanaka, K., Sasaki, M. S., Fujiwara, Y., and Akiba, H. (1977). DNA repair characteristics and skin cancers of xeroderma pigmentosum patients in Japan. *Cancer Res.* **37,** 490–495.

Tanaka, K., Naoyuki, M., Satokata, I., Miyamoto, I., Yoshida, M. C., Satoh, Y., Kondo, S., Yasui, A., Okayama, H., and Okada, Y. (1990). Analysis of a human DNA excision repair gene involved in group A xeroderma pigmentosum and containing a zinc-finger domain. *Nature* **348,** 73–76.

Tantin, D. (1998). RNA polymerase II elongation complexes containing the Cockayne syndrome group B protein interact with a molecular complex containing the transcription factor IIH components xeroderma pigmentosum B and p62. *J. Biol. Chem.* **273,** 27794–27799.

Tantin, D., Kansal, A., and Carey, M. (1997). Recruitment of the putative transcription-repair coupling factor CSB/ERCC6 to RNA polymerase II elongation complexes. *Mol. Cell. Biol.* **17,** 6803–6814.

Taylor, E. M., and Lehmann, A. R. (1998). Conservation of eukaryotic DNA repair mechanisms. *Int. J. Radiat. Biol.* **74,** 277–286.

Taylor, E., Broughton, B. C., Botta, E., Stefanini, M., Sarasin, A., Jaspers, N. G. J., Fawcett, H., Harcourt, S. A., Arlett, C. F., and Lehmann, A. R. (1997). Xeroderma pigmentosum and trichothiodystrophy are associated with different mutations in the *XPD* (*ERCC2*) repair/transcription gene. *Proc. Natl. Acad. Sci.* (*USA*) **94,** 8658–8663.

Tirode, F., Busso, D., Coin, F., and Egly, J. M. (1999). Reconstitution of the transcription factor TFIIH: Assignment of functions for the three enzymatic subunits, XPB, XPD, and cdk7. *Mol. Cell* **3,** 87–95.

Troelstra, C., van Gool, A., de Wit, J., Vermeulen, W., Bootsma, D., and Hoeijmakers, J. H. J. (1992). ERCC6, a member of a subfamily of putative helicases, is involved in Cockayne's syndrome and preferential repair of active genes. *Cell* **71,** 939–953.

van der Horst, G. T. J., van Steeg, H., Berg, R. J. W., van Gool, A. J., de Wit, J., Weeda, G., Morreau, H., Beems, R. B., van Kreijl, C. F., de Gruijl, F. R., Bootsma, D., and Hoeijmakers, J. H. J. (1997). Defective transcription-coupled repair in Cockayne syndrome B mice is associated with skin cancer predisposition. *Cell* **89,** 425–435.

van Gool, A. J., Verhage, R., Swagemakers, S. M. A., van de Putte, P., Brouwer, J., Troelstra, C., Bootsma, D., and Hoeijmakers, J. H. J. (1994). RAD26, the functional *S. cerevisiae* homolog of the cockayne syndrome B gene ERCC6. *EMBO J.* **13,** 5361–5369.

van Gool, A. J., Citterio, E., Rademakers, S., van Os, R., Vermeulen, W., Constantinou, A., Egly, J.-M., Bootsma, D., and Hoeijmakers, H. J. (1997a). The Cockayne syndrome B protein, involved in transcription-coupled DNA repair, resides in a RNA polymerase II containing complex. *EMBO J.* **16,** 5955–5965.

van Gool, A. J., van der Horst, G. T. J., Citterio, E., and Hoeijmakers, J. H. J. (1997b). Cockayne syndrome: Defective repair of transcription. *EMBO J.* **16,** 4155–4162.

van Hoffen, A., Natarajan, A. T., Mayne, L. V., van Zeeland, A. A., Mullenders, L. H. F., and Venema, J. (1993). Deficient repair of the transcribed strand of active genes in Cockayne's syndrome cells. *Nucleic Acids Res.* **21,** 5890–5895.

van Hoffen, A., Venema, J., Meschini, R., van Zeeland, A. A., and Mullenders, L. H. F. (1995). Transcription-coupled repair removes both cyclobutane pyrimidine dimers and 6-4 photoproducts with equal efficiency and in a sequenctial way from transcribed DNA in xeroderma pigmentosum group C fibroblasts. *EMBO J.* **14,** 360–367.

van Oosterwijk, M. F., Versteeg, A., Filon, R., van Zeeland, A. A., and Mullenders, L. H. F. (1996). The sensitivity of Cockayne's syndrome cells to DNA-damaging agents is not due to defective transcription-coupled repair of active genes. *Mol. Cell Biol.* **16,** 4436–4444.

Vermeulen, W., Scott, R. J., Potger, S., Muller, H. J., Cole, J., Arlett, C. F., Kleijer, W. J., Bootsma, D., Hoeijmakers, J. H. J., and Weeda, G. (1994a). Clinical heterogeneity within xeroderma pigmentosum associated with mutations in the DNA repair and transcription gene *ERCC3*. *Am. J. Hum. Genet.* **54,** 191–200.

Vermeulen, W., van Vuuren, A. J., Chipoulet, M., Schaeffer, L., Appeldoorn, E., Weeda, G., Jaspers, N. G. J., Priestley, A., Arlett, C. F., Lehmann, A. R., Stefanini, M., Mezzina, M., Sarasin, A., Bootsma, D., Egly, J.-M., and Hoeijmakers, J. H. J. (1994b). Three unusual repair deficiencies associated with transcription factor BTF2(TFIIH): Evidence for the existence of a transcription syndrome. *Cold Spring Harbor Symp. Quant. Biol.* **59,** 317–329.

Vuillaume, M., Daya-Grosjean, L., Vincens, P., Pennetier, J. L., Tarroux, P., Baret, A., Calvayrac, R., Taieb, A., and Sarasin, A. (1992). Striking differences in cellular catalase activity between two DNA repair-deficient diseases: Xeroderma pigmentosum and trichothiodystrophy. *Carcinogenesis* **13,** 321–328.

Vuillaume, M., Calvayrac, R., Best-Belpomme, M., Tarroux, P., Hubert, M., Decroix, Y., and Sarasin, A. (1986). Deficiency in catalase activity of xeroderma pigmentosum cell and simian virus 40-transformed human cell extracts. *Cancer Res.* **46,** 538–544.

Wakasugi, M., and Sancar, A., (1998). Assembly, subunit composition, and footprint of human DNA repair excision nuclease. *Proc. Natl. Acad. Sci. (USA)* **95,** 6669–6674.

Wang, Y.-C., Maher, V. M., Mitchell, D. L., and McCormick, J. J. (1993). Evidence from mutation spectra that the UV hypermutability of xeroderma pigmentosum variant cells reflects abnormal, error-prone replication on a template containing photoproducts. *Mol. Cell Biol.* **13,** 4276–4283.

Weeda, G., van Ham, R. C. A., Vermeulen, W., Bootsma, D., van der Eb, A. J., and Hoeijmakers, J. H. J. (1990). A presumed DNA helicase encoded by ERCC-3 is involved in the human repair disorders xeroderma pigmentosum and Cockayne's syndrome. *Cell* **62,** 777–791.

Weeda, G., Donker, I., de Wit, J., Morreau, H., Janssens, R., Vissers, C. J., Nigg, A., van Steeg, H., Bootsma, D., and Hoeijmakers, J. H. J. (1997a). Disruption of mouse *ERCC1* results in a novel repair syndrome with growth failure, nuclear abnormalities and senescence. *Curr. Biol.* **7,** 427–439.

Weeda, G., Eveno, E., Donker, I., Vermeulen, W., Chevallier-Lagente, O., Taieb, A., Stary, A., Hoeijmakers, J. H., Mezzina, M., and Sarasin, A. (1997b). A mutation in the XPB/ERCC3 DNA repair transcription gene, associated with trichothiodystrophy. *Am. J. Hum. Genet.* **60,** 320–329.

Wood, R. D. (1996). DNA repair in eukaryotes. *Annu. Rev. Biochem.* **65,** 135–167.

Wood, R. D. (1997). Nucleotide excision repair in mammalian cells. *J. Biol. Chem.* **272,** 23465–23468.

Yamaizumi, M., and Sugano, T. (1994). U.v.-induced nuclear accumulation of p53 is evoked through DNA damage of actively transcribed genes independent of the cell cycle. *Oncogene* **9,** 2775–2784.

4

Primary Immunodeficiency Mutation Databases

Mauno Vihinen*,[1,2] Francisco X. Arredondo-Vega,[3] Jean-Laurent Casanova,[4] Amos Etzioni,[5] Silvia Giliani,[6] Lennart Hammarström,[7] Michael S. Hershfield,[3] Paul G. Heyworth,[8] Amy P. Hsu,[9] Aleksi Lähdesmäki,[7] Ilkka Lappalainen,[1,10] Luigi D. Notarangelo,[6] Jennifer M. Puck,[9] Walter Reith,[11] Dirk Roos,[12] Richard F. Schumacher,[6] Klaus Schwarz,[13] Paolo Vezzoni,[14] Anna Villa,[14] Jouni Väliaho,[1] C. I. Edvard Smith[7,15,]*

[1] Institute of Medical Technology, FIN-33014 University of Tampere, Tampere Finland
[2] Tampere University Hospital, FIN-33520 Tampere, Finland
[3] Department of Medicine, Division of Rheumatology, Allergy, and Immunology, and Department of Biochemistry, Duke University Medical Center, Durham, North Carolina 27710, USA
[4] Unite Clinique d'Immunologie et d'Hématologie Pediatriques et Laboratoire INSERM U429, Hopital Necker-Enfants Malades, 75015 Paris, France
[5] Department Pediatrics, Rambam Medical Center, B. Rappaport School of Medicine, 31096 Haifa, Israel
[6] Istituto di Medicina Molecolare "Angelo Nocivelli," Department of Pediatrics, University of Brescia, I-25123 Brescia, Italy
[7] Division of Clinical Immunology, Karolinska Institute at Huddinge University Hospital, S-141 86 Huddinge, Sweden
[8] Department of Molecular and Experimental Medicine, The Scripps Research Institute, La Jolla, CA 92037, USA
[9] Genetics and Molecular Biology Branch, National Human Genome Research Institute, National Institutes of Health, Bethesda, MD 20892, USA
[10] Department of Biosciences, Division of Biochemistry, University of Helsinki, Finland
[11] Department of Genetics and Microbiology, University of Geneva Medical School, Ch-1211 Geneva 4, Switzerland

*Corresponding authors: Telephone: 46 8 608 9114. Fax: 46 8 774 55 38. E-mail: edvard.smith @cbt.ki.se

Advances in Genetics, Vol. 43

[12]Department of Experimental Immunohematology, Central Laboratory of the Netherlands Blood Transfusion Service, 1066 cx, Amsterdam, The Netherlands
[13]Department of Transfusion Medicine, University of Ulm, Ulm, Germany
[14]Department of Human Genome and Multifactorial Disease, Consiglio Nazionale delle Ricerche, Via Fratelli Cervi 93, 20090 Segrate (Milan), Italy
[15]Clinical Research Centre, Karolinska Institutet at Huddinge University Hospital, S-14186 Huddinge, Sweden

ABSTRACT

Primary immunodeficiencies are intrinsic defects of immune systems. Mutations in a large number of cellular functions can lead to impaired immune responses. More than 80 primary immunodeficiencies are known to date. During the last years genes for several of these disorders have been identified. Here, mutation information for 23 genes affected in 14 immunodefects is presented. The proteins produced are employed in widely diverse functions, such as signal transduction, cell surface receptors, nucleotide metabolism, gene diversification, transcription factors, and phagocytosis. Altogether, the genetic defect of 2140 families has been determined. Diseases with X-chromosomal origin constitute about 70% of all the cases, presumably due to full penetrance and because the single affected allele causes the phenotype. All types of mutations have been identified; missense mutations are the most common mutation type, and truncation is the most common effect on the protein level. Mutational hotspots in many disorders appear in C_pG dinucleotides. The mutation data for the majority of diseases are distributed on the Internet with a special database management system, MUTbase. Despite large numbers of mutations, it has not been possible to make genotype–phenotype correlations for many of the diseases.   © 2001 Academic Press.

I. INTRODUCTION

Primary immunodeficiency disorders (PIDs) affect the functioning of the immune system, because of mutations in genes involved in the production of components of immune response. Patients with these intrinsic defects have increased susceptibility to recurrent and persistent infections. PIDs were first identified 50 years ago (reviewed by Smith and Notarangelo, 1997) and have frequently served as prototype disorders. Thus, the first gene identified by positional cloning causes the X-linked form of chronic granulomatous disease (XCGD) (Royer-Pokora, et al., 1986). X-linked agammaglobulinemia (XLA) was the first human disorder in which a mutation in a cytoplasmic protein-tyrosine kinase was found (Tsukada et al., 1993; Vetrie et al., 1993), and the first inherited disorder in which gene therapy was attempted was adenosine deaminase (ADA) deficiency (Blaese et al., 1995; Bordignon et al., 1995). Furthermore, the first successful gene therapy trial in humans was recently reported for X-linked severe combined immunodeficiency (Cavazzana–Calvo et al., 2000).

PIDs do not differ from other genetic diseases with regard to the mutation spectrum. Because the immune system normally is not confronted with foreign intruders until after birth, PIDs do not result in prenatal death except when the affected gene also influences cells in organs other than the immune system.

Developmental defects may be seen in terms of organ size, but phenotypes due to increased infections normally appear weeks, months, or years after birth. Most immunodeficiencies are characterized by an increased susceptibility to infections. In brief, patients with B-cell and complement deficiency are prone to bacterial infections, whereas those with T-cell defects show an increased frequency to viral, parasitic, and fungal infections. Diseases affecting phagocytes mainly result in skin and oral bacterial infections and the formation of granulomas when the microorganisms spread to organs such as the liver. Fungal infections cause severe complications in these patients. Certain defects lead to an increased susceptibility to a few or even a single pathogenic agent(s). Thus, patients with X-linked lymphoproliferative syndrome (XLP), caused by mutations in the recently isolated *SH2D1A* gene (Coffey *et al.*, 1998; Nichols *et al.*, 1998; Sayos *et al.*, 1998), are selectively prone to Epstein-Barr virus infections, whereas patients with defects in interleukin 12 (IL-12) or interferon-γ signaling are sensitive to atypical forms of mycobacteria and also to *Salmonella*.

Some disorders of the immune system result in primarily autoimmune manifestations. Examples are mutations in the genes encoding mediators of apoptosis, such as *TNFSF6*, and *TNFRSF6* (Ochs *et al.*, 1999). To date, rather few patients belonging to this category have been reported, and therefore these diseases have not been included in this report. Certain primary immunodeficiencies result in an increased incidence of cancer. Examples are the X-linked disease, Wiskott-Aldrich syndrome (WAS), and the autosomal recessive disorder, ataxia-telangiectasia (A-T), both of which are included in this review. A-T also demonstrates that certain PIDs affect organs other than the immune system.

Genes for several of the more common forms of PIDs were identified in the 1990s (Ochs *et al.*, 1999, Smith and Notarangelo, 1997). Today more than 80 different PIDs are known, and the number is continuously increasing (Table 4.1). A recent example is the identification of mutations in the DNA methyltransferase gene, *DNMT3B*, in ICF syndrome (immunodeficiency, centromere instability, and facial anomalies). Due to the small number of identified cases (Hansen *et al.*, 1999; Xu *et al.*, 1999), this disease was not included in this chapter. The genes identified so far represent monogenic diseases, although the existence of modifier genes makes the transition between monogenic and polygenic diseases gradual.

The most common forms of PIDs are diseases with an X-linked recessive inheritance. Depending on whether the phenotype results in a selective survival disadvantage, carrier females may be at risk. As an example, in XLA, which shows a strong survival disadvantage of cells utilizing the affected X-chromosome, carrier females have never been found to develop symptoms, whereas in XCGD this is the case in females showing extreme lyonization.

Essentially all human genes will have been identified by the year 2000. Although these genes will also include those that cause PID, it will still take a long time before the corresponding patients have been identified, since many of these diseases are extremely rare.

Table 4.1. Immunodeficiencies and Affected Genes

Disease	Gene	Locus
Antibody deficiencies		
X-linked agammaglobulinemia (XLA)	BTK	Xq21.3-q22
IgA deficiency	IGAD1	6p21.3
Selective IgG subclass deficiencies	IGHG1	14q32.33
	IGHG2	14q32.33
	IGHG3	14q32.33
	IGHG4	14q32.33
κ-Chain deficiency	IGKC	2p12
Combined immunodeficiencies		
T⁻B⁻ SCID		
RAG-1	RAG1	11p13
RAG-2	RAG2	11p13
Adenosine deaminase deficiency	ADA	20q13.11
T⁻B⁺ SCID		
X-linked immunodeficiency with increased IgM (XHIM)	CD40LG	Xq26
X-linked severe combined immunodeficiency (XSCID)	IL2RG	Xq13
Jak3 deficiency	JAK3	19p13.1
Other CIDs		
Purine nucleoside phosphorylase deficiency	NP	14q13.1
ZAP-70 deficiency	ZAP70	2q12
CD3γ deficiency	CD3G	11q23
CD3ε deficiency	CD3E	11q23
TAP 2 peptide transporter deficiency	TAP2	6p21.3
MHCII deficiency (defect in CIITA)	MHC2TA	16p13
MHCII deficiency (defect in RFX5)	RFX5	1q21.1-q21.3
MHCII deficiency (defect in RFXAP)	RFXAP	13q14
MHCII deficiency (defect in RFXANK)	RFXANK	19p12
Other well-defined immunodeficiencies		
Wiskott-Aldrich syndrome (WAS)	WAS	Xp11.23-p11.22
DiGeorge syndrome	DGCR	22q11
Ataxia-telangiectasia	ATM	11q22.3
X-linked lymphoproliferative syndrome (XLP)	SH2D1A	Xq25
Familial hemophagocytic lymphohistiocytosis (FHL)	PRF1	10q21-22
ICF syndrome	DNMT3B	20q11.2
Phagocytic Immunodeficiencies		
Leukocyte adhesion deficiency 1 (LADI)	ITGB2	21q22.3
Chediak-Higashi syndrome	CHS1	1q42.1-q42.2
X-linked chronic granulomatous disease (XCGD)	CYBB	Xp21.1
Autosomal recessive CGD p22^{phox}	CYBA	16q24
Autosomal recessive CGD p47^{phox}	NCF1	7q11.23
Autosomal recessive CGD p67^{phox}	NCF2	1q25
IFN-γ receptor deficiency	IFNGR1	6q23-q24
	IFNGR2	21q22.1-q22.2
	IL12RB1	19p13.1
	IL12B	5q31.1-q33.1

Already today the number of mutations identified in unrelated families with primary immunodeficiency surpasses 2000. This has resulted in the development of mutation databases available on the Internet in order to handle the large amount of available information (Smith and Vihinen, 1996; Vihinen *et al.*, 1995). This chapter reviews mutations in 23 genes that cause 14 different PIDs. Since all the common forms of PIDs are included, as well as examples of more rare diseases, this chapter demonstrates the strength of international collaboration and how this can be integrated to serve as a useful source of bioinformations.

A. MUTbase System for the Storage and Analysis of Mutation Information

Immunodeficiency patient information has been collected in a large registry by the European Society for Immunodeficiencies (ESID) and in national patient registries. The ESID database contains clinical information about more than 8000 patients (Abedi *et al.*, 1995). The registry lists various immunodeficiency disorders, immunoglobulin values, therapy, and family history of patients. The ESID registry is available on the Internet at http://www.cnt.ki.se/esidregistry/.

The first immunodeficiency mutation database, BTKbase for XLA, was initiated in 1994 (Vihinen *et al.*, 1995, 1996a). Laboratories working on XLA patients were invited to join an international study group open to all investigators. Subsequently, similar databases have been generated for other well-defined immunodeficiencies. Previous reports of these registries were published in 1996 (Notarangelo *et al.*, 1996; Puck *et al.*, 1996; Roos *et al.*, 1996a; Schwarz *et al.*, 1996b; Vihinen *et al.*, 1996b, 1999b).

The MUTbase program suite provides easy, interactive, and quality-controlled submission of information to mutation databases (Riikonen and Vihinen, 1999). For further study of the databases on the World Wide Web, a number of tools are provided. The majority of the registries reported here have been generated and maintained using the MUTbase suite.

Mutation databases add to the knowledge of diseases and can be used in different ways, e.g., to analyze mutation mechanisms or to retrieve retrospective and prospective information on clinical presentation, disease phenotype, long-term prognosis, and efficacy of therapeutic options. The clinical information may be crucial for the development of new treatments including drug design.

A submission procedure was developed for easy access and use on the World Wide Web. This program writes the data in a format suitable for addition to a database, and performs several quality controls and warns of possible errors. The program calculates changes at the protein level by the mutation and checks for the numbering and type of the nucleotide(s) affected.

IDbases use the naming system developed by the Nomenclature working party of the Mutation Database Initiative (Antonarakis *et al.*, 1998). In addition, IDbases give a unique identifier, a Patient Identity Number (PIN). A PIN is a simple, comprehensive, and unambiguous name, which identifies the type and location of a mutation. Each entry has also an Accession number, a unique running number. The PIN consists of the type of mutation and a running number indicating mutations affecting the same amino acid or the same noncoding region. A more detailed description of PINs is given by Vihinen *et al.* (1999a).

The MUTbase distributes the registries on the World Wide Web, but it also generates from the available raw data new information, part of which is added to the entries, and it writes numerous interactive Web pages based on the performed analyses. Several of the pages produced can be used to retrieve entries from the databases.

The IDbases contain much information in addition to actual mutation data, making the registries valuable for physicians and scientists from many fields. ID registries report the submitting authors, submission date, and mutation information. It is possible to refer to database entries in publications, just as in the case of sequences not published elsewhere.

B. Genes and Proteins in PIDs

PID-causing genes can be found on a number of chromosomes (Table 4.1). X-chromosomal PIDs include XLA, HIGM1, XCGD, XLP, XSCID, and WAS. The majority of the genes code for multidomain proteins, but there are also single-domain proteins such as ADA and SH2D1A. The number of exons varies from 2 in RAG1 to 66 in ATM. Correspondingly, the proteins produced have lengths from 128 amino acids in SH2D1A to 3056 residues in ATM and molecular weights from 15 to 370 kDa. The largest databases are those for the X-chromosomal disorders XLA, XCGD, SCID, and WAS, and ATM of chromosome 11 origin.

Three-dimensional structure has been determined for domains of several PID-related proteins (Table 4.2). The PH domain, Btk motif, and SH3 domain structures have been determined for Btk (Hansson *et al.*, 1998; Hyvönen and Saraste, 1997), as well as the mouse ADA structure (Wilson and Quiocho, 1993), the extracellular fragment of human CD40 ligand (Karpusas *et al.*, 1995), and the SH2 domain structure of SH2D1A (Li *et al.*, 1999; Poy *et al.*, 1999). These structures as well as the model of the Btk SH2 and kinase domain (Vihinen *et al.*, 1994a, 1994b) can be used to interpret the protein structural consequences of the mutations. This information has been added to the BTKbase.

The functions of the PID-related proteins are very diverse in nature, indicating the importance of the numerous cellular functions for producing natural

Table 4.2. Immunodeficiency-Related Genes and Proteins

Disease	X-linked agammaglobulinemia (XLA)	X-linked lymphoproliferative syndrome	X-linked immunodeficiency with hyper-IgM
Prevalence	1/200,000	<1/1,000,000 live births in males	<1/1,000,000 live births in males
Gene name	*BTK*	*SH2D1A, (SAP, DSHP)*	*TNFSF5, CD40LG*
Protein name	Bruton agammaglobulinemia tyrosine kinase (BTK)	SH2D1A, SAP, DSHP	Tumor necrosis factor superfamily member 5, CD40 ligand (CD40L)
OMIM number	300300	308240	308230
Chromosomal location	Xq21.3	Xq25	Xq26
Sequence entry (DNA)[a]	U78027	HS1052M9	D31793–D31797
Number of exons	19	4	5
Total length of the gene (kb)	37.5	25	13
Sequence entry (mRNA)[a]	X58957	NM_002351 (GenBank); AL023651	X67878, Z15017
Length (bp)	2560	2500	1800
Sequence entry (amino acid)[b]	Q06187	060880	P29965
Length (amino acids)	659	128	261
Molecular weight of the protein (kDa)	77	15	33
3D structure (PDB entry)	1AWW; 1AWX; 1BTK; 1B55; 1BWN	1D1Z; 1D4T; 1D4W	1ALY
Internet address of the database	http://www.uta.fi/imt/ bioinfo/BTKbase/	http://www.uta.fi/imt/ bioinfo/SH2D1Abase/	http://www.uta.fi/imt/ bioinfo/CD40Lbase/

(*Continues*)

Table 4.2. (*Continued*)

Disease	X-linked severe combined immunodeficiency	JAK3-deficient severe combined immunodeficiency (autosomal recessive T⁻B⁺ SCID)	Adenosine deaminase deficiency	B-cell negative SCID, Omenn syndrome
Prevalence	1/50,000 to 1/100,000	<1/500,000 live births	$1/2 \times 100$ to 1×100 births	~1/100,000
Gene name	*IL2RG*	*JAK3*	*ADA*	*RAG1*
Protein name	γc or IL2Rγ	Janus kinase 3 (JAK3)	Adenosine deaminase (ADA)	Recombination activating protein 1 (RAG1)
OMIM number	300400, 308380	600173	102700	601457 (B cell negative SCID) 267700 (Omenn syndrome)
Chromosomal location	Xq13.1	19p12-13.1	20q13.11	11p13
Sequence entry (DNA)[a]	L19546	HSU70065, AC007201	M13792; (Wiginton et al., 1986)	
Number of exons	8	23	12	2
Total length of the gene (kb)	4	13.6	32	
Sequence entry (mRNA)[a]	NM_000206 (GenBank)	HSU09607	(Daddona et al., 1984 Wiginton et al., 1984)	M29474
Length (bp)	1451	3900	1500	6545
Sequence entry (amino acid)[b]	NP_000197.1	P52333	CAA26734; (Daddona et al., 1984)	P15918
Length (amino acids)	369	1124	363	1043
Molecular weight of the protein (kDa)	64	125	41	119
3D structure (PDB entry)			1ADD; 1FKW; 1FKX; 1UI0; 2ADA	
Internet address of the database	http://www.nhgri.nih.gov/ DIR/LGT/SCID/ IL2RGbase.html	http://www.uta.fi/imt/ bioinfo/JAK3base/	http://www.uta.fi/imt/ bioinfo/ADAbase/	RAG1:http:// www.uta.fi/imt/ bioinfo/RAG1base/

(*Continues*)

Table 4.2. (*Continued*)

Disease	B-cell negative SCID, Omenn syndrome	Wiskott-Aldrich syndrome (WAS) X-linked thrombocytopenia (XLT; THC1)	Ataxia-telangiectasia	Leukocyte adhesion deficiency type I
Prevalence	~1/100,000	Unknown	~1/100,000	1/100,000
Gene name	RAG2	WAS	ATM	ITGB2
Protein name	Recombination activating protein 2 (RAG2)	Wiskott-Aldrich syndrome protein (WASP)	ATM	β_2 subunit of β_2 integrin—CD18
OMIM number	601457 (B-cell-negative SCID) 267700 (Omenn Syndrome)	301 000 (WAS) 313 900 (XLT)	208900	116920
Chromosomal location	11p13	Xp11.23-p11.22 AH 007601; AF 196970	11q22-q23 U55702-U55757	21.922.3 X64071-X64083; X63924-X63926;
Sequence entry (DNA)[a]	3	12 9.5	66 150	16 2.8
Number of exons	M94633	NM_000377 (GenBank); U19927	NM_000051 (GenBank); U33841	M15395
Total length of the gene (kb)	5314	1806; 1796	9385	2247
Sequence entry (mRNA)[a]	P55895	NP_000368 (GenPept); P42768	AAC50298	P05107
Length (bp)	527	502	3056	769
Sequence entry (amino acid)[b]	59	53 (calculated MW) ~66 (apparent MW)	370	95
Length (amino acids)				
Molecular weight of the protein (kDa)				
3D structure (PDB entry)				
Internet address of the database	RAG2:http://www. uta.fi/imt/bioinfo/ RAG 2base/	—	http://www.vmresearch.org/ atm.htm and http://www.cnt.ki.se/ ATbase	http://www.uta.fi/imt/ bioinfo/ITGB2base/

(*Continues*)

Table 4.2. (*Continued*)

	MHC class II deficiency (bare lymphocyte syndrome)			
Disease	Defect in CIITA	Defect in RFXANK	Defect in RFX5	Defect in RFXAP
Prevalence	<1/2,000,000	<1/2,000,000	<1/2,000,000	<1/2,000,000
Gene name	MHC2TA	RFXANK	RFX5	RFXAP
Protein name	CIITA	RFXANK	RFX5	RFXAP
OMIM number	600005	603200	601863	601861
Chromosomal location	16p13	19p12 AD000812, AC003110	1q21.1-21.3	13q14
Sequence entry (DNA)[a]	20[c]	10 9	10[c]	3[c]
Number of exons	X74301	AF094760	X85786	Y12812
Total length of the gene (kb)	4534	1345	2262	2785
Sequence entry (mRNA)[a]	C2TA_HUMAN	RFXK_HUMAN	RFX5_HUMAN	RFXA_HUMAN
Length (bp)	P33076	O14593	P48382	O00287
Sequence entry (amino acid)[b]	1130	260	616	272
Length (amino acids)	123.5	28.1	65.3	28.2
Molecular weight of the protein (kDa)				
3D structure (PDB entry)				
Internet address of the database	—	—	—	—

(*Continues*)

Table 4.2. (*Continued*)

Disease	Chronic granulomatous disease			
	X-linked chronic granulomatous disease (X-CGD)	Autosomal CGD– p22^{phox} deficiency	Autosomal CGD– p47^{phox} deficiency	Autosomal CGD– p67^{phox} deficiency
Prevalence	1/250,000	<1/2,000,000	~1/500,000	<1/2,000,000
Gene name	CYBB	CYBA	NCF1	NCF2
Protein name	gp91^{phox}	22^{phox}	p47^{phox}	p67^{phox}
OMIM number	306400	233690	233700	233710
Chromosomal location	Xp21.1	16q24	7q11.23	1q25
Sequence entry (DNA)[a]	X05895	M61106; M62817; M61107; M62818	U57833-U57835;	U00776-U00788;
Number of exons	13	6	11	16
Total length of the gene (kb)	30	8.5	15	40
Sequence entry (mRNA)[a]	X04011	M21186, J03774	M25665, M26193	M32011
Length (bp)	4700	800	1400	2400
Sequence entry (amino acid)[b]	P04839	P13498	P14598	P19878
Length (amino acids)	570	195	390	526
Molecular weight of the protein (kDa)	91	22	47	67
3D structure (PDB entry)				
Internet address of the database	http://www.uta.fi/imt/ bioinfo/CYBBbase/	http://www.uta.fi/imt/ bioinfo/CYBAbase/	http://www.uta.fi/imt/ bioinfo/NCF1base/	http://www.uta.fi/imt/ bioinfo/NCF2base/

(*Continues*)

Table 4.2. (*Continued*)

	IFNγ-mediated immunity			
Disease	IFNγ1-receptor deficiency	IFNγ2-receptor deficiency	Interleukin-12 p40 deficiency	Interleukin-12 receptor β1 deficiency
Prevalence	<1/2,000,000	<1/2,000,000	<1/2,000,000	<1/2,000,000
Gene name	IFNGR1	IFNGR2	IL12RB1	IL12B
Protein name	Interferon-γ	interferon-γ	interleukin-12	interleukin-12 β-chain precursor
OMIM number	receptor α-chain precursor	receptor β-chain precursor	receptor β1-chain precursor	
Chromosomal location	107470	147569	601604	161561
Sequence entry (DNA)[a]	6q23-q24	21q22.1-q22.2	19p13.1	5q31.1-q33.1
Number of exons	AL0503377			
Total length of the gene (kb)	7	7		6
Sequence entry (mRNA)[a]	~50	33		
Length (bp)	J03143	U05875	U03187	M65272
Sequence entry (amino acid)[b]	2300	2200	2100	1400
Length (amino acids)	P15260	P38484	P42701	P29460
Molecular	489	337	662	328
weight of the protein (kDa)	55	38	73	37
3D structure (PDB entry)				
Internet address of the database	—	—	—	—

[a] EMBL sequence entry if not otherwise stated.
[b] SwissProt sequence entry if not otherwise stated.
[c] Number of exons in the human gene is predicted from the mouse gene.

and adaptive immune response. Thus, the proteins in PIDs are employed, e.g., in signal transduction, cell surface receptors, nucleotide metabolism, gene diversification, transcription factors, and phagocytosis. However, the exact function and mechanism of many of the proteins remain to be studied. As the activities suggest, the localization of these proteins ranges from the nucleus via cytoplasm and compartments to the cell membrane. Equally variable is the expression pattern of the proteins, although hematopoietic cells are prevalent.

A number of these proteins form complexes: RAG1 and RAG2, CIITA, RFXANK and RFXAP, gp91[phox], p22[phox], p47[phox], and p67[phox]. Mutations in successive steps of signal transduction pathways lead to PIDs, namely, IL2RG and Jak3 deficiencies to SCID and IL-12 and IFN-γ mutations to defects in IFN-γ mediated immunity.

A majority of the PIDs identified this far are recessive. However, in *IFNGR1* dominant mutations are also known. Consanguinity is common in families with autosomal recessive forms of immunodeficiencies.

II. ANTIBODY DEFICIENCIES

A. Mutations in X-Linked Agammaglobulinemia (BTKbase)

X-linked agammaglobulinemia (XLA) is a hereditary immunodeficiency caused by mutations in the gene coding for Bruton's agammaglobulinemia tyrosine kinase (Btk) (Tsukada *et al.*, 1993; Vetrie *et al.*, 1993). XLA is caused by a block in B-cell differentiation, resulting in severely decreased numbers of B lymphocytes and an almost complete lack of plasma cells, and negligible or very low immunoglobulin levels of all isotypes (Sideras and Smith, 1995). Patients have increased susceptibility to mainly bacterial infections because of virtually absent humoral immune responses. Among the most common clinical manifestations are upper and lower respiratory tract infections. Patients are treated with both antibiotics and immunoglobulin replacement therapy.

Btk is expressed in all hematopoietic lineages except for T lymphocytes and plasma cells (Smith *et al.*, 1994). The gene encoding murine Btk has also been cloned and sequenced (Rawlings *et al.*, 1993; Thomas *et al.*, 1993) (Table 4.2).

Btk belongs to the Tec family of related cytoplasmic protein tyrosine kinases (PTKs) formed by Tec, Itk/Tsk/Emt, and Bmx. They all consist of five distinct structural domains (Figure 4.1), from the N-terminus: a pleckstrin homology (PH) domain of about 120 amino acids, a Tec homology (TH) domain (about 60–80 residues), an Src homology 3 (SH3) domain of about 60 residues, an SH2 domain (about 100 amino acids), and the catalytic kinase domain of about 280 residues.

Btk is believed to be involved in transducing signals from the cell surface to the nucleus and has been shown to interact with several partners (for review, see Vihinen *et al.*, 1997a). The PH domain binds to phosphatidyl inositols and can function as a membrane localizing module. The TH domain contains a Zn^{2+}-binding motif (Hyvönen and Saraste, 1997; Vihinen *et al.*, 1997b) and a polyproline stretch. The SH2 and SH3 domains each recognize short peptide motifs bearing either phosphotyrosine (pTyr) residue or polyprolines, respectively, and link Btk to partner molecules. The kinase domain is the only catalytic region in Tec family kinases.

The three-dimensional structure has been determined for the PH domain and the first half of the TH domain (Hyvönen and Saraste, 1997) and for the SH3 domain (Hansson *et al.*, 1998). The structures of the other domains have been modeled (for reference, see Vihinen *et al.*, 1999b). Information concerning the consequences of all the mutations based on structural information is available in the BTKbase.

The disease-causing mutations and mutation types are indicated in Figure 4.1. Altogether there are 636 patients in the database. The patients represent 556 unrelated families (Table 4.3), and there are 401 unique mutations (72%). XLA mutations are scattered all along the *BTK* gene (Figure 4.1). The distribution of the mutations in the five structural domains is approximately according to the length of the domains, except for the TH domain. The most prevalent polymorphisms and variants affect positions 2031 (c/t), intron 8 at positions +70 (t/c) and 78 (a/g), as well as intron 14, position −29 (t/c).

Missense mutations comprise the most common mutational event, followed by nonsense and splice site mutations. Frameshift mutations constitute the largest group when deletions, insertions, and splice site mutations are added up. The missense mutations appear mainly in the first two positions within the codon. Nonsense mutations are relatively frequent, due mainly to mutations in the CpG dinucleotides. Substitutions in seven families alter the start codon and prevent expression of the protein.

The most frequently affected sites in point mutations are CpG dinucleotides. The 33 CpG dinucleotides in the coding region comprise only 3.3% of the *BTK* gene, although CG-to-TG or -CA mutations constitute 29.5% of all single base substitutions.

Intron mutations in 75 families cause aberrant splicing. Skipping of exon 9 causes an in-frame deletion in the C-terminus of the SH3 domain in two unrelated families. Insertions and deletions have been identified from 28 and 85 families, respectively. The larger deletions encompass multiple exons.

Many of the mutations affect functionally significant, conserved residues. The majority of the missense mutations in the PH domain are in the inositol compound-binding region. In the TH domain the missense mutations affect Zn^{2+} binding. No missense mutations are known to affect the SH3 domain. Most of the amino acid substitutions in the SH2 domain impair pTyr binding. In the kinase domain, the mutations are mainly on one face of the molecule, which is in charge of ATP, Mg^{2+}, and substrate binding. Although several missense mutations are expressed as protein (Holinski-Feder *et al.*, 1998), in many instances the mutated Btk becomes unstable (Conley *et al.*, 1998; Futatani *et al.*, 1998).

The disease is considered to have full penetrance. XLA in females with a skewed lyonization has not been reported, most likely due to the survival advantage of B lymphocytes utilizing the wild-type gene. Mutations have frequently

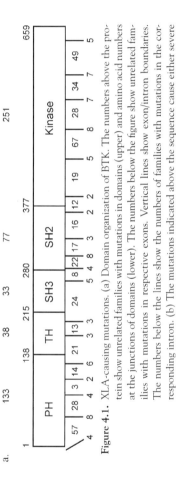

Figure 4.1. XLA-causing mutations. (a) Domain organization of BTK. The numbers above the protein show unrelated families with mutations in domains (upper) and amino acid numbers at the junctions of domains (lower). The numbers below the figure show unrelated families with mutations in respective exons. Vertical lines show exon/intron boundaries. The numbers below the lines show the numbers of families with mutations in the corresponding intron. (b) The mutations indicated above the sequence cause either severe (classical) or moderate XLA, whereas those denoted below the sequence are for clinically mild disease. @ insertion; # deletion; X nonsense mutation (stop codon).

118

b.

PH domain

```
                                P
                                L        N
                                H        C
      X   @ ##  @ PXFX  X  X  X  H  SP   #S#    ##    @ # #@# N# D  #  @  #  ##@ P  #           X              L
      # @ ##  @ PXFX  E  # # s RC       SP   #S#         @ # #@# N# D  #  @  #  ##@ P  #        ##@ P  #        L
  1   MAAVILESIF LKRSQQKKKT SPLNFKKRLF LLTVHKLSYY EYDFERGRRG SKKGSIDVEK ITCVETVVPE KNPPERQIP RGGEESSEME QISIIRFPY
      X                                    H
```

```
                X
      @#   #      XD#F P        X   H   #  #  #X
101   PFQVVYDEGP LYVFSPTEEL RKRWIHQLKN VIRNSDL
```

TH domain

```
        X          R                               P                      @
      X X   @   @ SG @  @             X       @#  @ P      ##@# #          X#          ##
139   VQ KYHPCFWIDG QYLCCSQTAK NAMGCQILEN RNGSLRPGSS HRKTKKFLPP TEEDQILKK PLPPEPAAAP VSTSE
```

SH3 domain

```
                    X
      #    X    @        ##X   @#  #      X   X    X  @  @   #  #          X
216   LKKVV ALYDYMPMNA NDIQLRKGDE YFILEESNLP WWRARDKNGQ EGYIPSNYVT EAEDSIEMYE
```

SH2 domain

```
          W                R          T                 X                                     R
          Q         P      E       GE         #### FA          # X       S  X   @   X  X       F    CQ PYF  FM  G  X
      X    X #      X #              GE        ##k## FA         # X       S  X   @   X  X       F    CQ PYF  FM  G  X
281   WYSKHMTRSQ AEQLLKQEGK EGGFIVRDSS KAGKYTVSVF AKSTGDPQGV IRHYVVCSTP QSQYYLAEKH LFSTIPELIN YHQHNSAGLI SRLKYPV
          W
```

Kinase domain

```
                   X              V                                                                           V        D
                   W              Y                                       R                                   XX  D   #R
      @X    #   #   G      #    # P @    LR   H   X   X  NE                                X  D    #       XX  D   #R
378   SQQ NKNAPSTAGL GYGSWEIDPK DLTFLKELGT GQFGVVRYGK WRGQYDVAIK MIKEGSMSED EFTEEAKVMM NLSHEKLVQL YGVCTKQRPI FIITEYM
```

```
                   X   V
                   W   Y   T         N  Q                                                                     W
                   R   R   I   XQ    X H E P S                              K                                 PX        N
      #    #    X  F V F D@   @P  X RQQ G ##GK#  #   # F # #  P G#  V        X  XS   ##L  #K P  #@    #@       N
478   ANG CLLNYLREMR HRFQTQQLLE MCKDVCEAME YLESKQFLHR DLAARNCLVN DQGVVKVSDF GLSRYVLDDE YTSSVGSKFP VRWSPPEVLM YSKFSSK
                   R                                                                  X
```

```
            K                         T
      V     G    R              F   S           @T        X       H  S
      Y   R#SR#  XD  XP  E  TC  X     @PD     XD  #  A    #P  G @#K  ##YX#@@   CH  L  R      #P
578   SDI WAFGVLMWEI YSLGRMPYER FTNSETAEHI AQGLRLYRPH LASEKVYTIM YSCWHEKADE RPTFKILLSN ILDVMDEES
            L              E                                          H        P
```

Figure 4.1. (*Continued*)

119

Table 4.3. Mutation Types (Numbers of Unrelated Families)

Gene or disease name	BTK	CD40LG	SH2D1A	IL2RG	JAK3	ADA
Missense	212	35	6	76	7	37
Nonsense	103	15	9	44	4	4
Deletion in frame	14	2		3	2	1
Deletion frameshift	86	14	5	32	0	4
Insertion in frame	3	0		1	0	
Insertion frameshift	32	10	1	9	1	
Splice site in frame	7	3		40	0	10[a]
Splice site frameshift	79	8	7	4	2	
Gross deletions	15	2	10	4		
Other	5		1	7	1	
Total	556	89	39	220	17	56

Gene or disease name	B⁻ SCID RAG1	B⁻ SCID RAG2	Omenn RAG1	Omenn RAG2	MHC2TA	RFXANK	RFX5	RFXAP	WAS	ATM
Missenese	2	2	6	2	1	1			32	42
Nonsense	3				1	3	2	2	25	47
Deletion in frame									1	26
Deletion frameshift				2		1		3	16	121
Insertion in frame									2	4
Insertion frameshift								1	13	41
Splice site in frame					3	1			4	84
Splice site frameshift						20	3		16	83
Gross deletions										1
Other									6	6
Total	5	2	6	4	5	26	5	6	115	455

Gene or disease name	CYBB	CYBA	NCF1	NCF2	IFNGR1	IFNGR2	IL12RB1	IL12B	ITGB2
Missense	103	6		4	3	1			14
Nonsense	102		3	2	1		3		
Deletion in frame	6			2	2				1
Deletion frameshift	45	1	45	3	5	1	1	1	3
Insertion in frame	4								1
Insertion frameshift	34	1		1	1				
Splice site in frame	46	2[a]		1	3[a]		1[a]		
Splice site frameshift	35			3					
Gross deletions	41	1		1					
Other									
Total	416	11	48	17	15	2	5	1	19

[a] Total of splice site mutations.

originated in the maternal grandfather or in male gametes coming from earlier generations (Conley et al., 1998).

The severity of XLA can be classified according to susceptibility to infections or based on decreased B-lymphocyte numbers and/or immunoglobulin levels. In certain kindreds the severity of XLA can vary among members carrying the same mutation, indicating that modifier genes may be involved. Most of the data in the BTKbase is for severe (classical) XLA. However, a high frequency of mutations resulting in mild XLA causes classical XLA in other families. Thus, at present there is no apparent genotype–phenotype correlation in XLA, although it seems likely that mutations having more subtle effects on enzyme function would be found primarily in mild disease.

III. T-CELL DEFICIENCIES

A. Mutations in X-Linked Lymphoproliferative Syndrome (SH2D1Abase)

Mutations in the gene coding for SH2D1A (also known as SAP and DSHP) have been found to cause X-linked lymphoproliferative syndrome (XLP, originally named Duncan disease, OMIM 308240). However, the molecular events behind XLP are still partially unclear, as SH2D1A defects have also been found in patients with atypical presentation and a substantial proportion of bona-fide XLP patients have no apparent defects in the gene (Brandau et al., 1999; Coffey et al., 1998; Nichols et al., 1998; Sayos et al., 1998; Yin et al., 1999).

XLP is a rare immunodeficiency characterized by extreme susceptibility to Epstein-Barr virus (EBV) infection (Purtilo et al., 1974; Schuster and Kreth, 1999). The overall prognosis in XLP is very poor, with over 70% of affected patients dying by 10 years of age (Seemayer et al., 1995). In affected male patients the infection by EBV leads to an uncontrolled expansion of EBV-infected B cells and cytotoxic T cells, often leading to lymphoma. The most common clinical manifestations include fulminant hepatitis, lymphoproliferative syndrome, dysgammaglobulinemia with low levels of IgG and high levels of IgM, aplastic anaemia, and vasculitis, variably associated. The major cause of death in these patients is fulminant hepatitis. The only effective therapy is bone marrow transplantation, with best results if performed prior to EBV infection. XLP patients are totally asymptomatic before EBV infection, with only minor immunological abnormalities (IgG subclasses deficiency) reported (Sullivan and Woda, 1989). For this reason it is of great importance to offer an early diagnosis to patients and families.

The XLP gene (SH2D1A) is divided into 4 exons spanning approximately 26 kb on genomic DNA (Table 4.2). It encodes a small protein of 128 amino acids that consists of a single SH2 domain (residues 6-102) followed by a

C-terminal extension (Coffey *et al.*, 1998; Nichols *et al.*, 1998; Sayos *et al.*, 1998). The majority of previously described SH2 domain proteins consist of additional signaling domains. Therefore, SH2D1A propably belongs to a novel family of small SH2 domain-containing proteins, formed by SH2D1A and a structurally similar EAT-2 (Coffey *et al.*, 1998; Poy *et al.*, 1999; Thompson *et al.*, 1996).

The SH2 domains bind phosphorylated tyrosines in a conserved manner. The specificity of each SH2 domain is achieved by recognition of 3–6 amino acids following the phosphotyrosine. The SH2 domain of SH2D1A has a unique specificity. It has also been shown to bind in a phosphorylation-independent manner and the specificity involves also residues N-terminal to phosphotyrosine (Li *et al.*, 1999; Sayos *et al.*, 1998).

SH2D1A is preferentially expressed in T lymphocytes and natural killer (NK) cells (Sayos *et al.*, 1998; Tangye *et al.*, 1999). In T lymphocytes, SH2D1A has been shown to bind the TIY^{281}AQV sequence in the cytoplasmic domain of SLAM (signal lymphocyte-activator molecule), regardless of the phosphorylation status of the tyrosine (Sayos *et al.*, 1998). The SLAM receptor is present on the surfaces of B and T cells and is involved in the activation of both types of cells (Cocks *et al.*, 1995). SH2D1A also binds *in vitro* to the cytoplasmic domain of an activating receptor of NK cells, 2B4, expressed mainly in T and NK cells (Matsuda *et al.*, 1990; Moran *et al.*, 1990; Tangye *et al.*, 1999). In both cases, the SH2D1A SH2 domain competes with the SHP-2 for binding to receptors. SH2D1A seems to inhibit recruitment of SHP-2 to the activated receptors or to prevent the phosphorylation of the receptors. In XLP patients, lack of inhibition of SHP-2 function could lead to inappropriate immune response following EBV infection.

The three-dimensional structure of the SH2D1A SH2 domain has been solved alone and complexed to phosphorylated or unphosphorylated SLAM-derived peptide (Li *et al.*, 1999; Poy *et al.*, 1999). In the unphosphorylated complex, ordered water molecules replace the interactions of the missing phosphate oxygens (Poy *et al.*, 1999). Specific interactions with N-terminal residues to phosphotyrosine increase the binding energy and allow recognition of the nonphosphorylated SLAM peptide (Li *et al.*, 1999; Poy *et al.*, 1999).

The SH2D1Abase includes 48 patients representing 39 unrelated XLP families (Figure 4.2) (Lappalainen *et al.*, 2000). The 21 unique mutations (53%) are spread along the entire gene, mainly confined to the SH2 domain. There are six different missense and three nonsense mutations. Four different deletions lead to frameshifts, and a number of gross deletions causes deletion of the whole gene. In five different families, the mutations are found in the intron sequences. In three cases mutations affect the initial Met (M1I, #M1X21) or the final stop codon (X129R). Moreover, one case of mutation in the 5'UT region in the putative CCAAT box has been also identified.

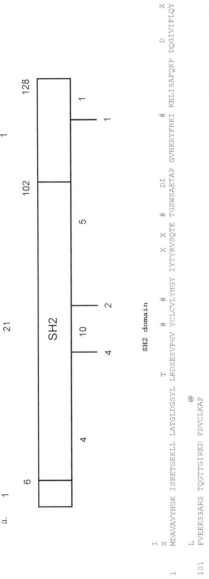

a.

6 4   21   SH2   102   1   128

4 10 2   5   1   1

b.

SH2 domain

```
     I
     X
  1  MDAVAVYHGK ISRETGEKLL LATGLDGSYL LRDSESVPGV YCLCVLYHGY IYTYRVSQTE TGSWSAETAP GVHKRYFRKI KNLISAFQKP DQGIVIPLQY
       T  #   #           X  X  #  DI                                                          D        X
         @
101  PVEKKSSARS TQGTTGIRED PDVCLKAP
       L
```

Figure 4.2. SH2D1A mutations. (a) Schematic representation of the SH2D1A protein structure and mutations. (b) Mutations in the SH2D1A protein. Notation as in Figure 4.1.

A rather large proportion of XLP patients (31%) have complete gene deletions. Nonsense mutations (21%) are more common mutational events compared to missense mutations (17%). This is due to the hotspot mutation, R55X, that is present in 6 unrelated families and is caused by C-to-T nucleotide change in the CpG dinucleotide. The other hotspot mutations are Q58X, occurring in two unrelated families, and a nucleotide change in intron 1, found in three different families.

Many of the missense mutations affect residues in the ligand-binding site of the SH2 domain. The R32T mutation disturbs the specific interactions required for phosphotyrosine recognition and binding. E67D and T68I mutations affect binding of the third residue following the phosphotyrosine. Three of the mutations most likely do not affect the structure of the SH2 domain, but highlight the importance of the short tail following the SH2 domain. Currently, the function of this C-terminal extension is unknown.

IV. COMBINED IMMUNODEFICIENCIES

Combined immunodeficiency, including severe combined immunodeficiency (SCID) is a deficit of both the cellular (T-lymphocyte) and humoral (antibody) arms of the immune system. Affected infants develop fatal infections unless reconstitution of the immune system is achieved, usually by bone marrow transplantation. There are several genetic forms of SCID.

A. Mutations in X-Linked Immunodeficiency with Hyper-IgM (CD40Lbase)

X-linked immunodeficiency with hyper-IgM (HIGM1) is an inherited combined immune deficiency caused by mutations in the gene coding for the CD40 ligand (CD40L), a molecule belonging to the superfamily of tumor necrosis factor (TNF) (Allen *et al.*, 1993; Aruffo *et al.*, 1993; DiSanto *et al.*, 1993; Fuleihan *et al.*, 1993; Korthäuer *et al.*, 1993). The gene has been recently renamed tumor necrosis factor superfamily member 5, or *TNFSF5* (Table 4.2). Failure of CD40L to functionally interact with its counter-receptor CD40 (expressed by B lymphocytes, dendritic cells, macrophages, and a variety of endothelial and epithelial cells) results in defective terminal B-cell maturation (with impaired immunoglobulin isotype switching) and defective responses to intracellular pathogens (Grewal and Flavell, 1996). Consequently, patients with HIGM1 show very low or undetectable serum IgG and IgA, with normal to elevated IgM, and have an increased susceptibility to bacterial and opportunistic infections (*Pneumocystis carinii* pneumonia and *Cryptosporidium*-related diarrhea in particular). Furthermore, they are at high risk for progressive liver disease (sclerosing cholangitis) and liver and intestinal tumors. Neutropenia is also a common manifestation. Treatment is based on regular use of intravenous immunoglobulins, prophylaxis of *Pneumocystis carinii* pneumonia with co-trimoxazole, and hygienic measures to prevent *Cryptosporidium* infections (Levy *et al.*, 1997). However, due to the severity of the disease (with less than 30% survival rate at 25 years of age), bone marrow transplantation from HLA-identical donors has recently been advocated.

The expression of CD40L is tightly regulated. In mature T cells, CD40L is mainly expressed by activated $CD4^+$ lymphocytes, with a different time-course depending on the activating agent. The expression of CD40L is reduced in newborns, possibly contributing to the physiological immune deficiency of this age, and is also inhibited by cyclosporin A (Brugnoni *et al.*, 1994; Fuleihan *et al.*, 1994).

CD40L is a type II transmembrane protein that contains a short intracytoplasmic tail (22 amino acids), a transmembrane domain (amino acids 23–46), and an extracellular TNF-homology domain (amino acids 123–261) (Figure 4.3). The extracellular domain immediately proximal to the transmembrane region contains a proteolytic cleavage site that is responsible for production of a soluble form of CD40L.

A three-dimensional structure of the extracellular region of CD40L has been determined by X-ray crystallography (Karpusas *et al.*, 1995). Similar to other members of the TNF superfamily, CD40L forms a homotrimer, with both hydrophobic and hydrophilic residues involved in the CD40-binding surface.

The *TNFSF5/CD40L* gene is organized in 5 exons (Shimadzu *et al.*, 1995; Villa *et al.*, 1994). Since 1994, mutations reported to the European Society for Immunodeficiencies have been collected in a database, CD40Lbase (Table 4.2) (Notarangelo *et al.*, 1996), which includes molecular, immunological, and clinical information. Altogether, 110 patients have been reported to the database, representing 89 unrelated families and 76 (85.4%) unique mutations (Table 4.3). Most of the mutations involve the extracellular region of the molecule (Figure 4.3).

Missense mutations comprise the most common mutational event; frameshift mutations (due to deletions or insertions) are the second most common type of mutations, when added up (Table 4.3). Some mutations occur in more than one family, and some residues are mutated to different codons in unrelated families. In particular, W140 is the most obvious mutational hotspot.

Among splice site mutations, some (mutations at the donor splice site of intron 4) allow for in-frame deletions, which, however, are functionally dead. Moreover, mutations at the last base of exons 1 and 2 have been identified that, though not resulting in amino acid changes (K52K, K96K), affect mRNA splicing and hence cause disease.

While some of the missense mutations affect the CD40-binding site (as for V126A and T147N), most of them disrupt CD40L function by other mechanisms (Bajorath *et al.*, 1995a, 1995b; Karpusas *et al.*, 1995). In particular,

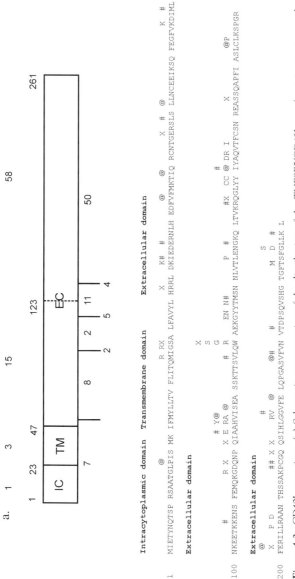

b.

```
Intracytoplasmic domain   Transmembrane domain        Extracellular domain
```

```
            @
1     MIETYNQTSP RSAATGLPIS MK IFMYLLTV FLITQMIGSA LFAVYL HRRL DKIEDERNLH EDFVFMKTIQ RCNTGERSLS LLNCEEIKSQ FEGFVKDIML
                                    R RX        X  K# # @        @       X # @                                K #
```

```
Extracellular domain
```

```
                                                   X
                                                   S
                                                   G
            #        R  X   X E RA @     # Y@    # R      EN N#      P #     #X  CC  @ DR I     X         @P
100   NKEETKKENS FEMQKGDQNP QIAAHVISEA SSKTTSVIQW AEKGYYTMSN NLVTLENGKQ LTVKRQGLYY IYAQVTFCSN REASSQAPFI ASLCLKSPGR
```

```
Extracellular domain
```

```
      @                         #                                          S
      X  P D      ## X X X   RV @     @#     #                             M D  #
200   FERILLRAAN THSSAKPCGQ QSIHLGGVFE LQPGASVFVN VTDPSQVSHG TGFTSFGLLK L
```

Figure 4.3. CD40L mutations. (a) Schematic representation of the distribution of the *TNFSF5/CD40L* gene/protein structure and mutations. The functional domains of the protein are shown as follows (IC, intracytoplasmic domain; TM, transmembrane domain; EC, extracellular domain). The last box (broken line) represents the tumor necrosis factor homology domain at the end of the EC domain. (b) Mutations in the CD40L. The tumor necrosis factor homology domain starts from residue 112. The K52K and K96K mutations are single-nucleotide substitutions that occur at the last base of exon 1 and 2, respectively, and interfere with mRNA splicing. Other notations as in Figure 4.1.

126

some mutations affect residues that participate in the generation of the hydrophobic core, and hence interfere with core packing and folding of the monomers (W140R, W140C, W140S, T254M), whereas other mutations involve buried residues at the interface between monomers, and thus prevent trimer formation (A123E, Y170C, G227V). The G38R mutation introduces a polar residue in the transmembrane domain, and thus affects expression of the protein at the cell surface.

In general, no strict genotype–phenotype correlation has been identified in HIGM1. Most patients have a severe phenotype, regardless of the type of mutations. However, rare patients with a mild phenotype (delayed onset, unusual presentation with parvovirus-related anemia) have been described, who carry mutations that result in generation of mutant proteins that retain the ability to bind CD40. In some cases, this is due to splicing defects that allow for residual production of wild-type mRNA (Seyama *et al.*, 1998). Finally, although carrier females are typically healthy (expression of functional CD40L on a proportion of cells due to random X-chromosome inactivation is sufficient to prevent disease), extreme lyonization may rarely result in hyper-IgM syndrome in carrier females (de Saint Basile *et al.*, 1999).

B. Mutations in X-Linked Severe Combined Immunodeficiency (IL2RGbase)

The most common form of SCID is X-linked (XSCID or SCIDX1, OMIM 300400) (Buckley *et al.*, 1997; Puck, 1999), accounting for approximately half of all SCID cases. XSCID is caused by mutations of the *IL2RG* gene (OMIM 308380). *IL2RG* encodes the common gamma chain (γc) of the receptors for interleukin-2 (IL-2) and several other cytokines. Cytokine receptor complexes containing functional γc are critical for lymphocyte differentiation and proliferation. The IL2RGbase catalogs 150 different *IL2RG* mutations in a total of 220 unrelated patients (Figure 4.4). XSCID-associated mutations occur throughout the 8 exons and splice sequences of *IL2RG* but with hotspots for mutation, and genotype–phenotype correlations are now recognized.

Infants with SCID develop severe infections in the first months of life as placentally acquired protective antibodies wane. By 3–6 months most boys with XSCID have failure to thrive, chronic diarrhea, and recurrent and persistent infections, often due to opportunistic pathogens, that fail to resolve with conventional treatment. T-cell counts are extremely low, and NK cells are also usually low, a phenotype of SCID that has been designated $T^- B^+ NK^-$. Lymphocyte functional studies reveal an absence of *in vitro* responses to mitogens and no specific antibody production by B cells. While XSCID is often suspected based on a family history of early male deaths in the maternal lineage, sporadic cases with newly arising mutations account for approximately one-third of cases.

a.

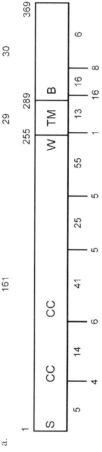

Figure 4.4. Mutations that cause XSCID. (a) Distribution of mutations throughout γc. S is signal sequence, CC denotes the conserved pairs of cysteines, W denotes the WSXWS motif, TM is the transmembrane region, and B is the Box 1/Box 2 domain. The numbers above the protein show amino acid numbers at the extracellular/transmembrane junction and at the transmembrane/intracellular junction. The uppermost numbers show families with mutations in those regions. (b) The mutations indicated above the sequence cause classical XSCID, while those below cause mild combined immunodeficiency, CID. In-frame insertions and deletions are indicated by underscored symbols. Complex mutations are represented by &, and mutations occurring in the splice regions are shown with a number in parentheses indicating unrelated cases. The conserved cysteines in exons 2 and 3 are underlined, as is the WSXWS motif. Other notations as in Figure 4.1.

b.

Signal sequence   **Conserved cysteines**

```
          V                                                      X        G        C
          K                       #                              G        K@       G      X   K#   C    #     E
          I                          (4)#                                             X    #  (6) X
  1  MLKPSLPFTS LLFLQLPLLG VG LNTTILTP NGNEDTTADF FLTTMPTDSL SVSTLPLPEV QCFVNVEYM NCTWNSSSEP QPTNLTLHYW YKNSDNDKVQ
                                                        N
```

Conserved cysteines   **Extracellular domain**

```
                                                                                        #
       Y  #C   X    DRX  X@  # P N  &   PFX     & X  P#P   (4) AN VS   R        P C # &  RS   XX     X   (6)
                                 X              X                   #          #        Q X      #
101  KQSHYLFSEE ITSGQQLQKK EIHLYQTFVV QLQDPREPRR QATQMLKIQN L VIPWAPENL TLHKLSESQL ELNWNNRFLN HCLEHLVQYR TDWDHSWTEQ
```

Extracellular domain   **Transmembrane domain**

```
           X   # # H    ## ## @  #                                              #
           X      C W CC  P YR# X R  C I@   X        R E##@   ##     (14)
                          ##   (1)#
201  SVDYRHKFSL PSVDGQKRYT FRVRSRFNPL CGSAQHWSEW SHPIHWGSNT SKEN PFLFAL EAVVISVGSM GLIISLLCVY FWLER
```

Box 1/Box 2 - Intracellular domain

```
                @   (7)
       X   # X#  X              #  X  X  #
           Q
286  TMPRI PTLKNLEDLV TEYHGNFSAW SGVSKGLAES LQPDYSERLC LVSEIPPKGG ALGEGPGASP CNQHSPYWAP PCYTLKPET
```

Figure 4.4. (*Continued*)

The *IL2RG* gene contains 8 exons (Table 4.2). The sequence contains a single Alu repeat in reverse orientation in intron 4. Characteristic of all members of cytokine receptor family, *IL2RG* codes for a signal sequence in exon 1; an N-terminal extracellular domain with two pairs of conserved cysteine residues; and a WSXWS motif implicated in cytokine binding. Box1/Box2 domain has similarity to the Src-related tyrosine kinases. Since the first compilation of *IL2RG* mutations in XSCID (Puck *et al.*, 1996) the total number of reported mutations has grown to 220.

The γc protein is widely expressed in hematopoietic cells in humans and mice and is found in the membrane receptors for multiple cytokines, including IL-2, IL-4, IL-7, IL-9, and IL-15 (Giri *et al.*, 1994; Kondo *et al.*, 1993; Noguchi *et al.*, 1993a; Russell *et al.*, 1993, 1994). Because the early differentiation of lymphocyte precursors depends on signaling through the IL-7 pathway and IL-2 is the major growth factor for T lymphocytes, γc has a critical role in lymphocyte differentiation and expansion. Upon cytokine binding, γc transmits an intracellular activation signal through its association with JAK3. Phosphorylation of JAK3 signals for the recruitment of one or more "signal transducer and activator of transcription" (STAT) proteins, specifically STAT5A and STAT5B, which dimerize and migrate into the nucleus to alter the cell's transcription program.

Canine (Henthorn *et al.*, 1994), murine (Cao *et al.*, 1993), and bovine (Yoo *et al.*, 1996) sequences show 88%, 81%, and 86% identity, respectively, to human *IL2RG* at the nucleotide level and 85%, 71%, and 82% identity of amino acids. Two mutations in canine *IL2RG* are known to cause XSCID.

Penetrance of XSCID is complete in males but, as a result of lyonization and negative selection against female lymphoid progenitors with mutant *IL2RG* on the active X chromosome, female carriers have normal immune function. Thus, female carriers have skewed X-inactivation patterns in their T and B cells, demonstrating the survival advantage for cells expressing a normal *IL2RG* allele.

In all but three cases reported to date, mutations in *IL2RG* severely compromise the function of γc and result in a severe phenotype. NK cells are usually absent, but have been found to be normal in XSCID patients; NK cell development is not mutation specific, as different patients with the same mutation have been shown to possess or to lack NK cells and *in vitro* NK cytotoxicity (Buckley *et al.*, 1999; Puck *et al.*, 1997). Mutations associated with XSCID are found throughout all 8 exons of *IL2RG* (Figure 4.4). Of the mutations in 220 unrelated patients, the most common are missense (76) and nonsense (44) point mutations; splice motifs are disrupted most often by point mutations (40), while deletions in splice sites have occurred in four locations (Table 4.3). The *IL2RG* exon mutations include 35 deletions and 10 insertions of 1 to 20 nucleotides as well as four large deletions of half the gene or more. Finally, there are 7 complex deletion/insertions.

The mutations in the *IL2RG* gene are not evenly distributed (Figure 4.4). Exon 1 contains four missense mutations, three altering the initial methionine and thus preventing protein translation, and the fourth changing the final base of the exon. The latter mutation, associated with a mild combined immunodeficiency (CID), produces a major disruption of splicing. The low amount of properly spliced mRNA contains a D39N amino acid substitution of unknown functional consequence (DiSanto *et al.*, 1994). Interestingly, only two missense mutations in exon 6 and none in exons 7 and 8 result in full-blown XSCID. The only missense mutation in exon 7 results in apparently normal amounts of γc protein with L293Q and occurred in a large kindred with mild X-linked CID (Russell *et al.*, 1994). Exons 2–through 5 have multiple missense mutations, all at residues conserved across species. In one patient with a T → C transition in exon 3, changing the fourth conserved cysteine to arginine, a spontaneous reversion of the mutation occurred, apparently in a lymphoid progenitor, resulting in development of a population of functional T cells and amelioration of his immunodeficiency (Stephan *et al.*, 1996).

The missense mutations in exon 5 lie within 20 residues that include the cytokine-binding homology region of the WSXWS motif. Based on sequence homology and crystal structure of gp130 (Bravo *et al.*, 1998) and IL4Rα (Hage *et al.*, 1999), this region comprises a conserved β sheet and the WSXWS loop, which has been implicated as a linker region between two domains and as part of the ligand-binding region in both gp130 and IL4Rα.

There are many mutations, either nonsense (29) or insertions and deletions with frameshifts (36), found in exons 1–6 of *IL2RG* which are proven or predicted to result in nonsense mediated decay of the mRNA (Cooper and Krawczak, 1993; Hentze and Kulozik, 1999), preventing protein translation. Splice site mutations (36) also result in unstable mRNA. Additionally, the mutations of the initial methionine result in failure to initiate translation. Thus 103 mutations (47%) apparently result in no protein production. Of the remaining families, 35% have missense mutations. Splice mutations in intron 7 cause inclusion of intron 7 in the final mRNA, which, when translated, introduces a frameshift. With the exception of the exon 7 missense mutation mentioned above, all mutations found in the intracellular portion of the gene result in truncation of the cytoplasmic tail and presumed or proven loss of downstream signaling to JAK3.

Several mutation hotspot sites are seen, five associated with CpG dinucleotides: 678 (5 mutations), 684 (7), 690–691 (18), all in exon 5, 868 (9) in exon 6, and 879 (9) in exon 7. Two other point mutations have been documented in four families each.

Three additional *IL2RG* regions are associated with multiple breakpoint mutations with small insertions or deletions. Slippage within a repeat of seven adenines in exon 3 accounts for five independent single-nucleotide insertions. A second site surrounding 833 in exon 6 has been mutated in 8 patients. Lastly, there are four large deletions, two of the whole gene, one missing the 5′ end, and the other missing the 3′ end. Both half-gene deletions have breakpoints in intron 4, which contains an Alu repeat, a recognized recombination site for large deletions.

One polymorphism has been reported in the coding region of *IL2RG*, a silent T instead of C at cDNA 1058; the frequency of this is less than 5%. Two common intronic polymorphisms are also recognized, a tetranucleotide in intron 1 which has either 3 or 4 repeats, and an a/g polymorphism in intron 2, 58 bases before exon 3. Neither of these appears to have functional consequences, since females homozygous for each polymorphism have no discernible phenotype.

C. Mutations in JAK3-Deficient T⁻ B⁺ SCID (JAK3base)

JAK3-deficient SCID is characterized by a lack of T (and NK) cells, with a normal to increased number of B cells in the periphery (T⁻ B⁺ SCID), leading to a profound impairment of both cellular and humoral immune responses (Anonymous, 1997). Recurrent severe respiratory infections, thrush, intractable diarrhea, and failure to thrive are the most common clinical manifestations in the first few months of life. Unless immune function is restored by successful bone marrow transplantation, the disease is usually fatal within the first years of life (Fischer *et al.*, 1990).

The disease is caused by mutations in the gene encoding JAK3, a member of the Janus associated kinase (JAK) family of nonreceptor protein tyrosine kinases, and has an autosomal recessive mode of inheritance (Candotti *et al.*, 1997; Macchi *et al.*, 1995; Russell *et al.*, 1995; Villa *et al.*, 1996).

Members of the JAK family, JAK1, JAK2, TYK2, and JAK3, show similar structure. They have a C-terminal kinase domain (JAK homology domain 1, JH1) flanked by a similar but nonidentical, noncatalytic pseudo-kinase domain (JH2) of supposed regulatory function (Figure 4.5). The N-terminal domains (JH6 and JH7) are necessary and sufficient for binding to the common γc, the additional domains (JH3–JH5) are thought to contribute to the *in vivo* assembly of the JAKs.

While the other members of the JAK family are ubiquitously expressed, JAK3 is preferentially (Gurniak and Berg, 1996; Kawamura *et al.*, 1994; Rane and Reddy, 1994), although not exclusively (Verbsky *et al.*, 1996), expressed in hematopoietic and lymphoid tissues. The genes encoding murine and rat JAK3 have also been cloned and sequenced (Takahashi and Shirasawa, 1994; Witthuhn *et al.*, 1994).

JAK3 is crucial for the signaling pathway of γc-containing interleukin receptors (namely, for IL-2, IL-4, IL-7, IL-9, and IL-15) that lack intrinsic catalytic activity (O'Shea, 1997). The intracellular JAK3 is always associated with another JAK protein that links to another (IL-specific) receptor subunit. The JAK kinases mediate cytokine-induced signal transduction by three sequential tyrosine phosphorylation steps that involve the cytokine receptor complex, the JAK proteins

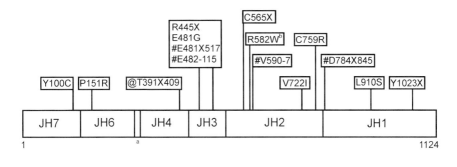

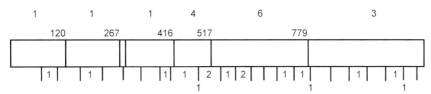

Figure 4.5. Schematic representation of the structure of the human *JAK3* gene and protein with the individual mutations on top. The JAK homology functional domains JH1–JH7 are shown as boxes (a denotes JH5 domain). The lower panel shows the distribution of mutations above the protein coding domains. The location of the mutations in the different exons, divided by lines indicating positions of introns, are shown below. One mutation (denoted b) causes two additional products, S568X581 and #S568-71.

themselves, and the STAT molecules, the secondary messengers that deliver the signal to specific DNA promoter sequences in the nucleus. Thus, JAK3 is essential for ontogeny and homeostasis of the lymphoid system.

There is evidence that dysregulation of JAK3 activity also may be involved directly or indirectly (Migone *et al.*, 1995; Tortolani *et al.*, 1995; Xu *et al.*, 1995) in various forms of hematopoietic malignancies (Danial *et al.*, 1995; Takemoto *et al.*, 1997).

Rapid regulation of JAK activity is achieved directly by differential autophosphorylation of tyrosine residues (Zhou *et al.*, 1997) in the putative activation loop of the kinase domain and by inhibitory proteins (Yoshimura, 1998). Indirect regulation via other molecules has also been described (Liu *et al.*, 1998).

The human JAK3 cDNA sequence was initially reported by Kawamura in 1994 (subsequently, 15 nucleotide corrections were made). The human *JAK3* gene maps to 19p12-13.1 (Hoffman *et al.*, 1997; Riedy *et al.*, 1996; Safford *et al.*, 1997) or 19p13.1–13.2 (Kumar *et al.*, 1996) within a cluster of proto-oncogenes and transcription factors and proximal to *TYK2*. The human *JAK3* gene is composed of 23 exons (Schumacher *et al.*, 2000) (Table 4.2). Prenatal diagnosis is available (Schumacher *et al.*, 1999).

The overall identity between the human and the murine JAK3 cDNA is 79% at the nucleotide level, and all the splice sites are located at identical positions.

The disease-causing mutations and mutation types are indicated in Figure 4.5. There are now 15 patients from 13 different families described (Table 4.3). One mutation was found in two siblings (#D784X845) and another (R445X) was detected in two related families and in a third unrelated family, indicating a possible hotspot or a founder effect. Given the small number of mutations detected, conclusions regarding distribution of the mutations in the seven structural domains must be drawn with caution; however, more than expected are clustered in the JH2 and JH3 domains.

Two polymorphisms in the noncoding region have been described: IVS7+13T → A and IVS13-30C → T.

Fourteen different events were point mutations. Seven of them were missense and three were nonsense mutations. Of the remaining, three were intronic splice site mutations, at least two of which caused aberrant splicing. One caused complete skipping of intron 16 (IVS16 + 2T → C), leading to a frameshift and a premature stop codon, and the other led to the use of a cryptic exonic splice site, again causing frameshift resulting in a premature stop codon (IVS9-2A → G).

Immunohistochemic studies showed no expression of JAK3 protein in the third patient with a splice site mutation (IVS21 + 2T → A; H. Gaspar, personal communication).

One of the missense mutations (C759R) results in constitutive phosphorylation of JAK3, which could not be unregulated by IL-2 stimulation *in vitro*. Another substitution in the JH2 domain (R582W) produced a mutated JAK3 protein that did not respond to IL-2 stimulation. The same genomic mutation also caused two other products, one with an in-frame deletion of 71 amino acids and the other characterized by a premature stop. The consequences of these and several other mutations are discussed in detail elsewhere (Bozzi *et al.*, 1998; Candotti *et al.*, 1997; Villa *et al.*, 1996).

D. Mutations in Adenosine Deaminase

Adenosine deaminase (ADA) is a monomeric zinc enzyme encoded by a 12-exon gene on chromosome 20q (Table 4.2). ADA deficiency is found in about 15% of all cases or 30–40% of patients with autosomal recessive inheritance of SCID. About 85–90% of ADA-deficient patients have SCID and are diagnosed by 1 year of age. About 15% have milder immunodeficiency and are diagnosed later in the first decade ("delayed onset"), or in the second to fourth decades ("late/adult onset"). Screening has identified several healthy individuals with low or absent erythrocyte ADA activity, but higher activity in nucleated cells ("partial ADA deficiency"). Among all SCID patients, those with ADA deficiency have the most profound lymphopenia, involving T, B, and NK cells (Buckley *et al.*, 1997). Nonlymphoid organ dysfunction, presumably related to the purine metabolic disorder, occurs variably, including a growth abnormality of costochondral junctions causing flaring of rib ends on X-ray, various kinds of CNS dysfunction, and elevated hepatic transaminases and occasionally overt hepatitis. In addition to bone marrow transplantation, ADA deficiency can be treated by enzyme replacement therapy with polyethylene glycol-modified bovine ADA (PEG-ADA, Adagen®, Enzon, Inc.), as well as with a combination of PEG-ADA and gene therapy (Hershfield, 1998; Hershfield and Mitchell, 1995; Onodera *et al.*, 1999).

In humans, the highest ADA expression occurs in the cytoplasm of lymphoid cells, particularly immature thymocytes. ADA apparently protects these cells from toxic effects of its substrates adenosine and deoxyadenosine, large amounts of which are generated by the active cell turnover that occurs in the thymus associated with T-cell maturation, and in lymph nodes during the response to antigen (reviewed by Hershfield and Mitchell, 1995). Various effects of adenosine and deoxyadenosine, and of their metabolites, contribute to the clinical manifestations of ADA deficiency (Hershfield and Mitchell, 1995). Two effects of deoxyadenosine are evident in red cells of immunodeficient patients: an increased content of deoxyadenosine nucleotides (dAXP, primarily dATP), and inactivation of the enzyme S-adenosylhomocysteine (AdoHcy) hydrolase. Red cell dAXP (dATP) elevation is greatest in patients with SCID and minimal in healthy individuals with "partial" ADA deficiency. In nucleated cells, dATP pool expansion can interfere with DNA repair and replication, and can induce apoptosis. This may be the primary mechanism underlying T lymphopenia. AdoHcy accumulation can interfere with cellular methylation reactions. Other toxic effects of adenosine and deoxyadenosine have been described, and adenosine acting via G-protein-coupled cell surface receptors may influence aspects of lymphocyte responses to antigens.

The entire human ADA gene has been sequenced (Wiginton *et al.*, 1986). The crystal structure of the closely related mouse ADA shows a parallel α/β barrel architecture with 8 central β strands and 8 peripheral α helixes. A zinc ion essential for activity is situated in the active site pocket coordinated with three histidine and one aspartate residues (Wilson *et al.*, 1991). By analogy with the structure of mouse ADA, only a few of the known ADA missense mutations alter residues at the active site, including E217, which binds substrate, and H15 and H17, which coordinate with the zinc ion. Most missense mutations are located in peripheral helices distant from the active site. Where studied, low levels of immunoreactive ADA protein have been found in cell lines of patients with SCID, suggesting that many missense mutations reduce protein stability or expression.

ADA deficiency is genetically heterogeneous. Sixty mutations (63% missense, 17% splice site, 14% deletions, and 7% nonsense mutations) have

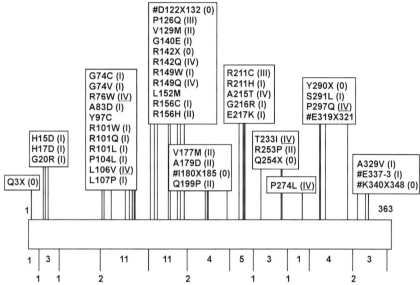

Figure 4.6. Mutations that cause ADA deficiency. Three large 5′ genomic deletions are not shown (two of 3.25 kb that include the promoter and exon 1, and *a* > 30-kb deletion that includes exons 1–5). For exonic mutations, the number in parentheses after the mutation name denotes the effect on ADA activity, as defined in Arredondo-Vega *et al.* (1998): deletion and nonsense mutations are designated "0," and missense mutations as "I," "II," "III," or "IV" based on increasing ADA activity derived from constitutively expressing an ADA cDNA carrying the mutation in an ADA deletion strain of *E. coli.* Patients with genotypes composed only of alleles from group "0" or "I" were very likely to have SCID, whereas those with at least one allele from group "II" or "III" were very likely to have a milder "delayed-onset" or "late/adult onset" phenotype; individuals who possessed one group "IV" allele were healthy and had "partial ADA deficiency" (Arredondo-Vega *et al.*, 1998). Mutations designated IV were also associated with "partial" deficiency and found to provide substantial residual activity by semiquantitative methods (Hirschhorn *et al.*, 1990).

been identified (Table 4.3, Figure 4.6). One study group that included 50 unre-lated and phenotypically heterogeneous patients comprised 43 genotypes derived from 42 different mutations, half of which occurred in single families; 6 muta-tions accounted for 48% of the 100 alleles: R211H (13%), G216R (11%), 955–959delGAAGA (7%) [referred to elsewhere as 955(1050)del5 or #E319X321], R156H (6%), A329V (6%), and L107P (5%) (Arredondo-Vega *et al.*, 1998).

There is evidence of a good correlation between genotype and both clinical and metabolic phenotype. Certain missense and some splicing mutations found in healthy individuals with "partial" deficiency and in immunodeficient patients with delayed or late-onset phenotypes have not been found in patients with SCID, the most common phenotype (Hirschhorn *et al.*, 1990; Santisteban *et al.*, 1993). We have quantitated the ADA activity expressed from cloned ADA cDNAs bearing 29 patient-derived missense mutations in an *Escherichia coli* strain with a deletion of the homologous bacterial ADA gene (Arredondo-Vega *et al.*, 1998). Expressed ADA activity, which ranged over 5 orders of magnitude, was correlated with clinical and metabolic findings of 49 immunodeficient and 3 healthy "partially deficient" individuals whose 42 different mutations included 28 of the expressed alleles (Arredondo-Vega *et al.*, 1998). Of 31 SCID patients, 27 carried mutant alleles that together expressed <0.05% of wild-type ADA activity, whereas 19 of 21 patients with milder phenotypes expressed >0.05% activity. Erythrocyte dAXP content correlated inversely with the level of expressed ADA activity (Arredondo-Vega *et al.*, 1998).

Apparent exceptions to the correlation between genotype and phenotype emphasize the role of particular genetic mechanisms in determining phenotype. For example, a discordant phenotype in ADA-deficient siblings was attributable to different efficiencies of producing normal ADA mRNA from an allele with a 5′ splice donor site mutation (Arredondo-Vega *et al.*, 1994). Two ADA-deficient patients who underwent spontaneous clinical remissions were discovered to show mosaicism (Hirschhorn *et al.*, 1994, 1996).

E. Mutations in B⁻ SCID and Omenn Syndrome (RAGbases)

B⁻ SCID and Omenn syndrome (OS) are caused by mutations in either of the two recombination-activating genes, *RAG1* and *RAG2* (Schwarz *et al.*, 1996a; Villa *et al.*, 1998). However, the fact that OS is caused by defects of the same gene(s) as B⁻ SCID came as a surprise, as the two diseases look quite different at first glance.

The B⁻ SCID patients exhibit no mature T or B cells, and show a complete absence of lymph nodes and tonsils. The clinical presentation is characterized by infections starting in the second or third month after birth. The patients show opportunistic infections, chronic persistent disease of the airways, and local and systemic bacterial infections. The recurrent infections lead to a failure to thrive, and EBV and CMV virus infections can cause lethal complications.

In addition to a clinical picture of severe immunodeficiency, OS presents some peculiar aspects, such as an early onset of a severe and generalized exudative erythrodermia, lymphoadenopathy and hepatosplenomegaly, hypereosinophilia, and elevated IgE levels. OS patients have a variable number of circulating T cells, poorly responsive to *in vitro* stimulation by mitogens and antigens. The analysis of the T-cell repertoire in OS patients reveals the presence of a restricted oligoclonality, already present at the thymic level (Signorini *et al.*, 1999; Villa *et al.*, 1998). In spite of the high IgE levels, B cells are absent or highly reduced in their number, with consequent low serum levels of immunoglobulins IgA and

IgG. The remarkable hypereosinophilia and the increased IgE serum levels suggest skewing toward a Th2 phenotype, as supported by the predominant secretion of IL-4 and IL-5 and a decreased production of the Th1 cytokine, IL-2, and IFN-γ.

While the conservative treatment of B⁻ SCID and OS patients consists of antibiotics and supportive measures, bone marrow transplantation represents the only curative tool.

RAG1 and *RAG2* are the first players in the V(D)J recombination process and were cloned because of their ability to rearrange artificial V(D)J recombination substrates in a fibroblast cell line (Oettinger *et al.*, 1990; Schatz *et al.*, 1989). Both human genes are located on chromosome 11 in a tail-to-tail configuration separated by 10.3 kb (Table 4.2). The coding regions contain a single exon, but one intron is located in the 5′UTR of *RAG1* and two introns interrupt the 5′ UTR of *RAG2*. RAG proteins are expressed in the early stages of lymphocyte development, and their expression is tightly regulated during the cell cycle. Knockout mutations in either *Rag* gene lead to severe combined immunodeficiency in mice, with a complete block of B- and T-cell development, while NK cells are spared (Mombaerts *et al.*, 1992; Shinkai *et al.*, 1992).

V(D)J recombination is directed by signal DNA sequences named recombination signal sequences (RSSs), which flanking each receptor gene segment and consist of a conserved heptamer and a nonamer separated by a nonconserved spacer of either 12 or 23 nucleotides. *RAG1* recognizes and binds the nonamer sequence and, after the recruitment of *RAG2*, the complex *RAG1/RAG2* introduces a double-strand break exactly at the border between a coding element and the heptamer of a RSS.

The RAG genes have a role not only in the first endonucleolytic step of the V(D)J recombination process but also in the following steps of DNA-end processing (Besmer *et al.*, 1998). Santagata *et al.* (1999) have reported a 3′ flap endonuclease activity of the RAG proteins and suggested a direct role in generating junctional diversity during the recombination process in a physiological setting. In addition, an *in vitro* transposase activity has been shown in RAG proteins (Agrawal *et al.*, 1998; Hiom *et al.*, 1998).

Two biologically active core regions, encompassing amino acids 384–1008 of *RAG1* and 1–387 of *RAG2*, have been defined on the basis of the ability of these truncated proteins to carry out recombination of episomal plasmid substrates *in vitro*. Within the RAG1 "core" segment, amino acids 376–477 are crucial for DNA binding and the region is related to the homeodomain of the Hin invertase family, whose members recognize nonamer-like sequences. The region also contributes to the heterodimerization domain, allowing the interaction with *RAG2* (Figure 4.7). The N-terminal portion of *RAG1* is involved in the interaction with the nuclear transport protein Srp1 through three basic regions, and its removal impairs the recombination process.

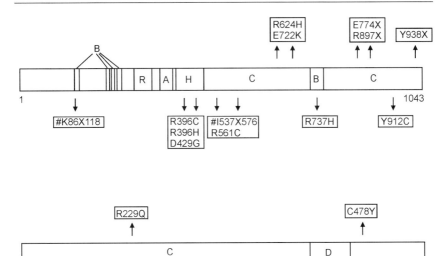

Figure 4.7. Schematic representation of the full-length human RAG1 (top) and RAG2 (bottom). Mutations identified in B⁻ SCID and in Omenn syndrome patients are indicated above and below the diagram of the protein, respectively. The RAG1 protein contains Basic domains I, IIa, IIb, III (B), a ring-finger domain (R), a zinc-finger domain A (A), a homeodomain (H), a core domain (C), and a zinc-finger domain B (B). The RAG2 protein contains a core domain (C) and an acidic domain (D).

The active core of RAG2 spans amino acids 1–387 and shows a probable sixfold repeated motif of 50 amino acids related to the kelch/mipp motif (Callebaut and Mornon, 1998) (Figure 4.7). The remaining part of the protein seems to play a yet undefined role in efficient recombination of Ig heavy-chain segments (Kirch et al., 1998).

About 30% of B⁻ SCID and more than 75% of OS show abnormalities in RAG genes as summarized in Figure 4.7. Prevalent polymorphic amino acids affect A156V, R244H, and R249H in RAG1 and V8I in RAG2.

Nowadays 5 of 14 B⁻ SCID patients have been reported, three with RAG1 and two with RAG2 mutations (Schwarz et al., 1996a) (Table 4.3). Among the RAG1 patients, one was homozygous for a stop mutation, while the two others were compound heterozygotes for a nonsense and a missense substitution. One of the RAG2 patients was homozygous for a missense mutation, while the other patient had a missense change on one allele and a deletion of the entire locus on the other, encompassing both RAG1 and RAG2. All these mutants showed less than 1% of recombination activity on artificial extrachromosomal V(D)J recombination substrates, suggesting that B⁻ SCID alleles are null alleles.

On the contrary, OS patients seem to maintain partial *RAG* activity. To date, seven cases have been reported, six with mutations in *RAG1* and one in *RAG2*. Most *RAG1* mutations affect specific regions of the proteins, and all patients carry at least one missense mutation (Figure 4.7). In fact, some of the missense mutations affect the homeodomain region of *RAG1* responsible for the recognition and binding to the nonamer motif of the RSS, while others map to the region involved in the interaction with RAG2 or regions responsible for the nuclear localization. The patient with a *RAG2* defect bears two missense mutations located in the active core on the same side of the propeller, both at the end of the second–third loop (Callebaut and Mornon, 1998).

Biochemical analysis of the Omenn syndrome mutants for DNA binding to RSS, cleavage activity, and the ability to interact with *RAG2* has revealed a decreased but not absent activity of the proteins. Likewise, all these mutants showed a reduced but not abolished recombination activity with extrachromosomal substrates in reconstituted cells. These data support the hypothesis that a partial defect in the ability to assemble a stable RAG/DNA complex can lead to OS. The residual activity of the proteins allows a limited number of recombination events to occur, giving rise to the development of oligoclonal T lymphocytes whose abnormalities are responsible for some clinical features of the syndrome. In contrast, severe impairment due to the complete absence of activity of RAG protein is responsible for B⁻ SCID. The analysis of a large spectrum of B⁻ SCID and OS cases will allow us to better understand the correlation between genotype and phenotype.

F. Major Histocompatibility Complex Class II Deficiency

Major histocompatibility complex class II Deficiency (MHCII deficiency), also referred to as the bare lymphocyte syndrome, is a rare autosomal recessive immunodeficiency disease resulting from the absence of MHCII expression at the surface of all cells that should normally display them (reviewed by Reith *et al.*, 1999b). These include epithelial cells in the thymus, professional antigen-presenting cells (APCs) such as dendritic cells, B cells, and macrophages, and nonprofessional APCs induced with interferon-γ. Consequently, the function of the adaptive immune system is severely crippled, both because positive selection of CD4$^+$ thymocytes is perturbed and because the ability to present antigenic peptides to CD4$^+$ T cells is lost. The resulting absence of CD4$^+$ T-helper cell activation leads to a severe impairment of humoral and cellular immune responses to foreign antigens.

Clinical manifestations include primarily septicemia and recurrent infections of the gastrointestinal, pulmonary, upper respiratory, and urinary tracts (reviewed by Reith *et al.*, 1999b). Patients are prone to bacterial, fungal, viral, and protozoal infections. Infections start within the first year of life. Subsequent evolution of the disease is characterized by a progression of the infectious complications until death ensues, generally before the age of 5 years. Almost all patients exhibit repeated and severe intestinal infections, protracted diarrhea, malabsorption, and failure to thrive.

The optimal symptomatic care available consists of the prophylactic use of antibodies capable of penetrating cells, intravenous administration of immunoglobulins, and parental nutrition. These means reduce the frequency and severity of the clinical problems, but can ultimately not prevent progressive organ dysfunction and death. The only curative treatment currently available is bone marrow transplantation, which has been reported to have a relatively poor success rate in this disease (reviewed by Reith *et al.*, 1999b).

The inability to express MHCII molecules is due to a deficiency in the transcription of MHCII genes. This results from mutations in trans-acting regulatory genes encoding transcription factors required for the activation of MHCII promoters (reviewed by DeSandro *et al.*, 1999; Reith *et al.*, 1999a, 1999b). The disease is genetically heterogeneous. Patients have been assigned to four "complementation" groups (A, B, C, and D). These groups reflect the existence of mutations in four trans-acting regulatory genes that are essential and highly specific for expression of MHCII genes (reviewed by Boss, 1999; Fontes *et al.*, 1999; Pan-Yun Ting and Zhu, 1999; Reith *et al.*, 1999a, 1999b). The *MHC2TA* gene mutated in patients from group A encodes a non-DNA-binding co-activator protein called CIITA (MHC class II trans-activator) (Steimle *et al.*, 1995) (Table 4.2). The genes that are mutated in patients from groups B (*RFXANK*, also called *RFX-B*), C (*RFX5*), and D (*RFXAP*) encode three different subunits of RFX (regulatory factor X), a heterotrimeric DNA-binding complex that binds to the X box cis-acting sequence of MHCII promoters (Durand *et al.*, 1997; Masternak *et al.*, 1998; Nagarajan *et al.*, 1999; Steimle *et al.*, 1995). RFX binds cooperatively with two other transcription factors (X2BP or CREB, and NF-Y) that recognize specific sequences within MHCII promoters (reviewed by Boss, 1999; Reith *et al.*, 1999a, 1999b). CIITA is tethered to MHCII promoters via multiple protein–protein interactions with the DNA-bound multicomponent complex formed by binding of RFX, X2BP (CREB), and NF-Y (Masternak *et al.*, submitted).

The genes for RFX subunits are expressed in a wide variety of cell types, including MHCII-negative cells. In contrast, the gene encoding CIITA exhibits a highly regulated expression profile that dictates the cell type specificity, induction, and level of MHCII expression (reviewed by Boss, 1999; Fontes *et al.*, 1999; Pan-Yun Ting and Zhu, 1999; Reith *et al.*, 1999a, 1999b). *MHC2TA* thus functions as a master control gene for MHCII expression. Transcription of MHC2TA is controlled by three independent promoters exhibiting different cell type specificity and responsiveness to induction by interferon-γ (Muhlethaler-Mottet *et al.*, 1997, 1998).

Several functional domains have been identified in CIITA (Figure 4.8A) (reviewed by Boss, 1999; Fontes *et al.*, 1999; Pan-Yun Ting and Zhu, 1999; Reith

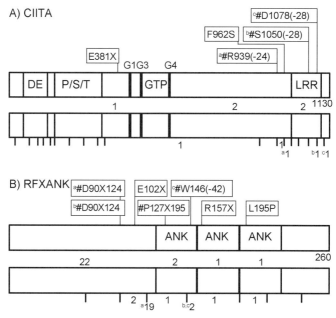

Figure 4.8. Molecular defects in the four MHC class II deficiency complementation groups. For each factor, the upper map shows the positions of domains and known mutations within the protein. The lower maps indicate the positions of introns (vertical lines below) and the distribution of mutations within the domains (above) or within the introns and exons (below). For CIITA, RFX5, and RFXAP, positions of the introns are predicted in part from the corresponding mouse genes. (A) Domains indicated in CIITA are an acidic region (DE), a proline/serine/threonine-rich region (P/S/T), three sequences (G1, G3, G4) constituting a GTP-binding motif, and a region containing four leucine-rich repeats (LRR). The three in-frame deletions ([a]#, [b]#, [c]#) are the result of splice mutations leading to exon skipping; sizes in amino acids of the deleted exons are indicated in parentheses. (B) Mutations in RFXANK (also called RFX-B). The three ankyrin repeats (ANK) are shown. Three of the deletions ([a]#, [b]#, [c]#) result from splice site mutations leading to out-of-frame exon skipping. Deletions [a]# and [b]# result from mutations at different splice sites. Mutation [a]# has been found in 19 unrelated patients. (C) Mutations in RFX5. The DNA-binding domain (DBD) and a proline-rich region (P) are indicated. The three out-of-frame deletions ([a]#, [b]#, [c]#) result from splice site mutations leading to the use of a displaced "cryptic" splice site. (D) Mutations in RFXAP. Positions of an acidic region (DE), a glutamine-rich region (Q), and a putative nuclear localization signal (NLS) are indicated.

et al., 1999a, 1999b). The N-terminus contains a segment that is rich in acidic amino acids (DE), and three proline/serine/threonine-rich regions (P/S/T), that are believed to function as transcription activation domains. The central portion of the protein contains three sequences (G1, G3, and G4) constituting a

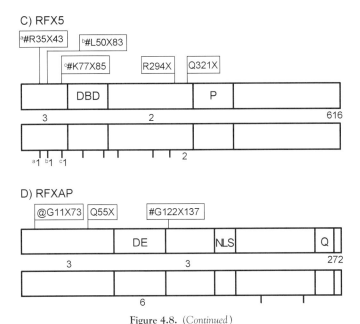

Figure 4.8. (*Continued*)

GTP-binding motif believed to play a role in nuclear import (Harton *et al.*, 1999). Finally, the C-terminus contains a putative protein–protein interaction domain consisting of four leucine-rich repeats (LRR). Five different mutations have been identified (Bontron *et al.*, 1997; Peijnenburg *et al.*, 2000; Quan *et al.*, 1999; Steimle *et al.*, 1993) (Table 4.3). One is a nonsense mutation leading to a truncated protein lacking the last two-thirds of the protein. The remaining are a missense mutation, and three splice site mutations entailing the skipping of in frame exons, in the vicinity of the LRR region.

The only recognizable motif in RFXANK is a protein–protein interaction domain consisting of three ankyrin (ANK) repeats (Masternak *et al.*, 1998; Nagarajan *et al.*, 1999) (Figure 4.8B). Structures have been determined for the ankyrin repeat regions of several other proteins. Mutations in RFXANK represent the most frequent molecular defect in MHCII deficiency: group B accounts for almost 70% of the patients that have been classified into the four complementation groups (Fondaneche *et al.*, 1998). Seven different mutations have been identified in 26 unrelated patients (DeSandro *et al.*, 1999; Masternak *et al.*, 1998; Nagarajan *et al.*, 1999; Lisowska-Grospierre *et al.*, submitted) (Table 4.3). Nineteen patients have the same mutation (#D90X124), indicating the existence of a founder effect (Lisowska-Grospierre *et al.*, submitted). All mutations affect the ankyrin repeats. One is a missense mutation lying within the third repeat. The remaining mutations are nonsense or splice site mutations leading to proteins lacking all or part of the ankyrin repeat region.

RFX5 (Figure 4.8C) belongs to a family of DNA-binding proteins sharing a well-characterized 74-amino acid DNA-binding domain (DBD) called the RFX motif (Emery *et al.*, 1996; Steimle *et al.*, 1995). The structure of the DBD of another member of the RFX family (RFX1) has been determined. It adopts a structure of the winged helix–turn–helix type, but contacts the DNA in an unusual and novel fashion (Gajiwala *et al.*, 2000). RFX5 also contains a proline-rich region that is reminiscent of certain transcription activation domains. Five different mutations have been defined (Peijnenburg *et al.*, 1999a, 1999b; Steimle *et al.*, 1995; Villard *et al.*, 1997a). All are nonsense or splice site mutations (Table 4.3), leading to the synthesis of severely truncated proteins lacking the DBD and/or the C-terminal moiety containing the proline-rich region.

RFXAP (Figure 4.8D) contains acidic- (DE) and glutamine-rich (Q) regions resembling activation domains (Durand *et al.*, 1997). There is also a putative bipartite nuclear localization signal (NLS). Only three different mutations have been described in 6 unrelated patients (Durand *et al.*, 1997; Fondaneche *et al.*, 1998; Villard *et al.*, 1997b). All mutations lead to the synthesis of severely truncated proteins lacking the DE region and/or the NLS and Q regions.

The most frequent origin of the MHCII-deficiency patients is the Mediterranean basin (North Africa, Turkey, Spain) (Fondaneche *et al.*, 1998). In the large majority of cases (39/42), the patients are homozygous for the mutation. In several cases, the same mutation has been found in more than one unrelated patient. These observations are consistent with the fact that the majority of patients have a family history involving documented or suspected consanguinity (Fondaneche *et al.*, 1998).

Despite the genetic heterogeneity in the cause of the disease, MHCII deficiency represents a remarkably homogeneous phenotypic entity. No characteristic features discriminating between patients from different complementation groups have been described. However, the analysis of CIITA and RFX5 knockout mice suggests that phenotypic differences between patients from different groups may exist (Chang *et al.*, 1996; Clausen *et al.*, 1998; Reith *et al.*, 1999a; Williams *et al.*, 1998). These mouse models exhibit two clear differences. First, there is a clear difference in the pattern of residual MHCII expression detected in certain cell types. Second, RFX5 knockout mice display strongly reduced levels of MHC class I expression in certain cell types (B and T cells), while this is not observed in CIITA knockout mice (Reith *et al.*, 1999a). In the human disease, differences in residual MHCII expression and a variable effect on MHC class I expression have both been observed, but no clear correlation with the genetic defect has yet been established (reviewed by Reith *et al.*, 1999b).

V. OTHER WELL-DEFINED IMMUNODEFICIENCIES

A. Mutations in Wiskott-Aldrich Syndrome and X-Linked Thrombocytopenia

Wiskott-Aldrich syndrome (WAS) is a primary immunodeficiency due to mutations in the gene encoding the Wiskott-Aldrich syndrome protein (WASP)
(Derry *et al.*, 1994). WAS is caused by a signal transduction defect in hematopoietic cells that, through dysorganization of the cytoskeleton, may result in increased
spontaneous apoptosis of blood cells (Rawlings *et al.*, 1999). Characteristic abnormalities include thrombocytopenia with a small platelet volume, lymphopenia
(increasing with age), lymphocyte depletion in the T-dependent pericortical areas
of lymphnodes and in the thymus, and abnormal T- and B-cell functions. Patients
with WAS exhibit a defective delayed type of hypersensitivity, lack isohemagglu-
tinins, fail to produce antibodies after immunization with linear polysaccharides,
and respond poorly to certain protein antigens. The clinical presentation is characterized by petechiae and bloody diarrhea, recurrent upper and lower respiratory
infections, eczema, autoimmune manifestations, and an increased risk of malignancies (Ochs, 1998). The clinical and immunological causes of the disease vary
widely from one patient to another. Treatment modalities include antibiotics and
supportive measures. The only definitive therapy available is hematopoietic stem
cell transplantation.

    X-linked thrombocytopenia (XLT) can be distinguished from classical
WAS since its definition includes thrombocytopenia with small platelets, absence
of or only mild and transient eczema, normal susceptibility to infections, and
sometimes subclinical immune dysfunction (Ochs, 1998).

    The human *WASP* gene locates to the X-chromosome at Xp11.23-p11.22
and was identified through positional cloning (Derry *et al.*, 1994). It encompasses
9 kb, codes for 12 exons, and encodes a cytoplasmic scaffold protein of 502 amino
acids (Table 4.2). WASP is regulated by two promotors (Hagemann and Kwan,
1999; Petrella *et al.*, 1998) and is expressed in all hematopoietic lineages including
early progenitors (Parolini *et al.*, 1997; Stewart *et al.*, 1996). The murine *Wasp*
gene is highly conserved and was artificially inactivated to create WAS animal
models (Derry *et al.*, 1995; Snapper *et al.*, 1998; Zhang *et al.*, 1999).

    In humans, the WASP family includes WASP itself, WASF1-3 (SCAR1-
3), and WASL (N-WASP) (Bi and Zigmond, 1999). Common to all members of
the WASP family is a Polyproline-rich region (PP) (about 90 residues) followed
by a Verprolin homology domain (V) (15–20 amino acids) and a Cofilin homology region (C) (15–20 residues) (Figure 4.9). These domains are bracketed by
a Basic domain (B) on the amino-terminal side and by an Acidic domain (A)
at the carboxy-terminal end. In addition, both, WASP and WASL, encompass a

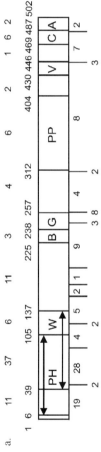

Figure 4.9. WAS-causing mutations. (a) Domain organization of WASP. The numbers above the protein show amino acid numbers at the junctions of domains. The numbers below the protein show unrelated families with mutations (upper) in respective exons (lower). Vertical lines show exon/intron boundaries. The numbers below the lines show the numbers of families with mutations in the corresponding intron. The protein contains Pleckstrin homology domain (PH), WASP homology domain (W), Basic domain (B), GTPase-binding domain (G), Polyproline domain (PP), Cofilin homology domain, and Acidic domain (A). (b) Mutations in the WASP protein. Notation as in Figure 4.1.

b.

PH and WH domain

```
                                                         C
                                                         S
           @ XX@        T   # ## K X#   P  X  M  ED    X  V       P      SM G @#  #XL H XD     C
1    MSGGPMGGRP GGRGAPAVQQ NIPSTLLQDH ENQRLFEMLG KKCLTLIATAV VQLYLALPPG AEHWTKEHCG AVCFVKDNPQ KSYFIRLYGL QAGRLLWEQE
```

PH and WH domain   **AA 137-224**

```
                                                   @
     S     @#                 R       K #    @    # #
101  LYSQLVYSTP TPFFHTFAGD DCQAGLNFAD EDEAQAF RAL VQEKIQKRNQ RQSGDRRQLP PPPTPANEER RGGLPLPLH PGGDQGGPPV GPLSLGLATV
```

AA 137-224   **Basic domain**   **GTPase binding domain**

```
          X              X
     X                   #           G
201  DIQNPDITSS RYRGLPAPGP SPAD KKRSGK KKISKAD IGA PSGFKHVSHV GWDPQNG
```

AA 258-311

```
               X      #    X
          X                 X
258  FDV NNLDPDLRSL FSRAGISEAQ LTDAETSKLI YDFIEDQGGL EAVRQEMRRQ E
```

Polyproline domain

```
     X     @         @        #       #
312  PLPPPPPPS RGGNQLPRPP IVGGNKGRSG PLPPVPLGIA PPPPTPRGPP PPGRGGPPPP PPPATGRSGP LPPPPPGAGG PMPPPPPPP PPP
```

AA 405-429   **Verprolin homology** **AA 447-468**   **Cofilin homology**   **Acidic domain**
                   domain   domain

```
                                                                  N
     @         #                         @          @@    K@     @      #   #      @
405  SSGNGP APPPLPPALV PAGGLAPGG G RGALLDQIRQ GIQLNK TPGA PESSALQPPP QSSEGLVG AL MHVMQKRSRA IHSSDE GEDQ AGDEDEDDEW DD
```

Figure 4.9. (*Continued*)

Pleckstrin homology domain (PH) at their N-terminus (about 120–140 residues) and a consensus sequence, known as GBD, which binds GTPases of the Rho family.

The precise function of WASP in mammalian cells is uncertain, but since WAS lymphocytes, platelets, granulocytes, and monocytes are defective in regulating the cortical actin cytoskeleton, it has been suggested that WASP is involved in actin polymerization and actin cytoskeleton organization (Remold-O'Donnell, *et al.*, 1996). WASP has been shown to associate with numerous partners including tyrosine kinases (e.g., Btk, Itk, Tec, Fyn, c-Src), adaptor proteins (Nck and Grb2), phospholipase $C\gamma_1$, proteins involved in actin polymerization (WIP, PST-PIP, Arp 2/3 complex), and proteins controlling actin cytoskeleton reorganization (Cdc42, Rac).

WASP may function as scaffold on a pathway connecting cell surface receptors—either serpentine receptors acting via heterotrimeric G proteins or receptor tyrosine kinases—to the actin cytoskeleton (Bi and Zigmond, 1999). After the activated GTP-bound Rho family member Cdc42 interacts directly or indirectly with WASP, the intramolecular interaction of the Basic and Acidic domains is interrupted. The altered WASP structure allows co-localization of monomeric actin (Verprolin domain) and the Arp2/3 complex, which localizes where polymerized F-actin is in a dynamic state (Acidic domain), possibly enhancing the latter's F-actin nucleation activity (Higgs *et al.*, 1999; Machesky and Insall, 1998).

No complete three-dimensional WASP structure has been derived, yet the solution structure of the WASP-GBD domain bound to Cdc42 was recently resolved (Abdul-Manan *et al.*, 1999). In addition, the EVH1 domain of Enabled (Ena), which is closely homologous to the WASP-WH1 domain, was crystallized and classified as a distinct member of the PH domain superfamily (Prehoda *et al.*, 1999). This now offers the opportunity to assess the influence of mutations on the WASP structure.

The disease-causing mutations submitted to WASPbase and the mutation types are indicated in Figure 4.9. There are 115 unrelated WASP families representing 148 patients, accounting for 85 (74%) unique mutations (Table 4.3). Three gross deletions (exons 1–2, exons 1–7, exons 8–12) have been detected; a smaller deletion exhibits the loss of the last nucleotide of exon 6 and the adjoint 11 nucleotides of intron 6. There is one splice defect in intron 11 leaving open the putative splice acceptor, and 1 mutation of the stop codon TGA. Twenty splice site mutations as well as 89 mutations affecting the coding region have been reported.

The WASP mutations affecting the coding region are unevenly distributed along the *WASP* gene. The PH and WH1 domain together account for 76% of all mutations but for only 26% of all amino acids. In addition, while the Cofilin homology region harbors 8.5% of the mutations and comprises only 4% of the WASP amino acids, the Polyprolin domain accounts for 8.5% of the mutants and resembles 18% of the protein. There are no mutations in the GBD and Verprolin homology domains.

The most common mutational events are missense mutations, which are mostly concentrated in the first third of the protein, followed by nonsense and splice site mutations. About 40% of all mutations cause a frameshift, when deletions, insertions, and splice site mutations are grouped together. Nonsense mutations are frequently due to alterations of CpG dinucleotides.

The most frequent point mutations are at CpG dinucleotides. Although 54 CpG nucleotides encompass only about 3.6% of the coding nucleotides, CG-to-TG or -CA changes constitute 35% of the mutations in the coding region. Three unique mutations, all located at CpG doublets (R41X, R86G or H or S, R211X) were seen more than five times. The most common site affected is R86 (9 events). Intron mutations in 20 different families cause aberrant splicing, with 4 splice events predicted to be in-frame. Deletions and insertions have been detected in 21 and 15 families, respectively.

WAS is considered to have full penetrance. A WASP defective female with a skewed lyonization has been reported to exhibit clinical features of WAS, most likely due to a disturbed X-inactivation program (Parolini et al., 1998). WASP mutations are frequently inherited and originated in the maternal grandfather or earlier male generations.

The clinical severity of WAS can be classified, but does not correlate with WASP mutations, but rather with the expression level of wild-type WASP protein (Ochs, 1998). XLT patients and milder WAS forms tend to have substitutions in the N-terminus of WASP, while severe WASP is mostly caused by loss of expression mutants. Since patients within one family or patients of different kindreds with identical mutations can present a wide spectrum of clinical symptoms, parameters other than the WASP mutations, such as genetic background or epigenetic events, may influence the WASP phenotype.

B. Mutations in Ataxia-Telangiectasia (ATbase)

Ataxia-telangiectasia (A-T; Louis-Bar syndrome) is a rare multisystem autosomal recessive disorder caused by mutations in the ATM gene. The disease is characterized by cerebellar ataxia, conjuctival telangiectasias, immunodeficiency, sensitivity to radiomimetic agents, and cancer predisposition (Savitsky et al., 1995a). It is also believed that there is an increased risk for tumors in the heterozygous carriers of the ATM gene, especially breast cancer and certain leukaemias (Easton, 1994; Vořechovský et al., 1997). The immunodeficiency affects both humoral and cellular immune systems. IgA-, IgE-, and IgG2 deficiencies are characteristic of the disease, and patients are prone to infections, primarily sinopulmonary infections. Patients are therefore treated with immunoglobulin replacement and antibiotics, but the disease is still often lethal at a young age.

The gene responsible for A-T was identified by positional cloning and designated *ATM* (ataxia-telangiectasia mutated) (Savitsky *et al.*, 1995a, 1995b) (Table 4.2). ATM is expressed in all tissues and cell types studied, with minor species of various sizes, possibly representing splice variants. The deduced amino acid sequence revealed similarities to the catalytic domain of the p110 subunit of phosphatidyl-inositol-3-kinases at the C-terminal part of *ATM* (Figure 4.10) and to a growing family of *ATM*-related genes that are implicated in cell cycle regulation, control of telomere length, and/or response to DNA damage (Savitsky *et al.*, 1995a, 1995b). The ATM protein interacts with Abl, firmly linking it to control of cell division and differentiation (Shafman *et al.*, 1997). The ATM protein has also been found to bind p53 directly and is responsible for its serine 15 phosphorylation, thereby contributing to the activation and stabilisation of p53 during ionizing radiation-induced DNA damage (Khanna *et al.*, 1998).

The mutations described thus far in the *ATM* gene are indicated in Figure 4.10. The mutations are scattered along the gene, in all exons, in most of the introns, and even in the promoter region (Castellví-Bel *et al.*, 1999). Frameshift deletions are the most common mutations, followed by mutations causing defective splicing. There are quite a few polymorphisms/variants found in the gene. The mutation data is from an Ataxia-Telangiectasia Mutation Database created by P. Concannon and R. A. Gatti, and ATbase.

Almost 300 different mutations have been identified along the gene (Table 4.3), most of which are unique to single families. However, for some mutations there is evidence of a founder effect in selected ethnic populations (Norwegian, Costa Rican, Moroccan Jewish, Polish, and British patients) (Telatar *et al.*, 1998). The majority of mutations are predicted to result in protein truncation and give rise to the classic phenotype of A-T (Concannon and Gatti, 1997), but other mutations can also give rise to the classic phenotype of the disease. There are, however, patients with milder manifestations, referred to as "A-T variants." The mutations in these cases are usually either very small in-frame deletions or leaky splice site mutations which allow correct splicing part of the time, leading to considerably reduced, but not absent, levels of the ATM protein (Gilad *et al.*, 1998).

IgA deficiency is a common feature of patients with A-T (Vořechovský, *et al.*, 1999). Allelic heterogeneity in A-T patients does not seem to account for the lack of IgA, since the same *ATM* mutation is found in IgA-deficient and IgA-proficient patients in the same family. The notion of an A-T-cell-mediated B-cell differentiation defect may be supported by a correlation of IgA levels and T-cell defects seen in some of the A-T patients. A growth delay in A-T patients may theoretically contribute to the IgA deficiency, and this idea is supported by the reduced levels of IgA found in patients with genetic defects associated with other forms of developmental retardation. Analysis of the patients in the ATbase, though, suggests only a weak positive correlation ($r = 0.35$) of age and serum IgA, indicating that the age factor alone cannot account for the difference between IgA-deficient and IgA-proficient A-T patients.

Strict genotype–phenotype correlations regarding immunodeficiency are difficult to establish, as completely different immunological phenotypes can be seen in siblings with the same mutations and in ATM-deficient and -proficient mice. In three Swedish sibpairs with A-T, both siblings in the first sib pair had IgA below the detection limit of 0.07 g/liter, but one had a total absence of IgG whereas the other showed only slightly reduced levels of the IgG2 subclass. In the second sibpair, both siblings had IgG2 deficiency but only one of them had IgA deficiency. Finally, in the third pair, one showed normal Ig levels whereas the other had IgG2 deficiency. However, these differences might be due to leaky splice site mutations that may allow correct splicing to varying degrees in the siblings and thereby perhaps contribute to the Ig deficiencies in only one of them. This is not the case, however, in at least one of the above families, where the siblings have a homozygous frameshift insertion. In the other two families this could be the reason, because only one mutation is known (in one family there is a splice site mutation) (Laake et al., unpublished). Homozygous ATM-deficient knockout mice show lower serum IgG1, IgG2a, IgG2b, IgG3, and IgA levels than their ATM-proficient or heterozygous littermates, but the variation between homozygous littermates in isotype levels is quite extensive (Xu et al., 1996; Y. Gu, personal communication). These findings suggest that background genes and/or environmental factors may play an important role in the development of the A-T phenotype.

VI. PHAGOCYTIC IMMUNODEFICIENCIES

A. Mutations in X-Linked Chronic Granulomatous Disease (CYBBbase)

X-linked chronic granulomatous disease (XCGD) is a hereditary immunodeficiency caused by mutations in CYBB, the gene coding for the 91-kDa glycoprotein of the phagocyte oxidase (gp91^{phox}) (Dinauer et al., 1987; Roos et al., 1996a, 1996b; Royer-Pokora et al., 1986; Teahan et al., 1987) (Table 4.2). XCGD is caused by the lack of superoxide generation in phagocytic leukocytes (neutrophils, eosinophils, and monocytes/macrophages), resulting in a decreased ability of these cells to kill ingested pathogenic microorganisms (Quie et al., 1967). As a result, patients have increased susceptibility to bacterial and fungal infections. Among the most common clinical manifestations are infections of the lungs, the gastrointestinal tract, the skin, and the lymph nodes (Roos and Curnutte, 1999). Patients are treated with intracellularly active antibiotics and with γ-interferon.

gp91^{phox} is exclusively expressed in phagocytic leukocytes, where it forms a heterodimer with p22^{phox}. gp91^{phox} contains two hemes and one FAD

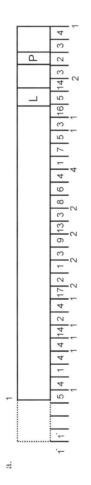

Figure 4.10. A-T-causing mutations. (a) The boxes show the domain organization of ATM (L, leucin zipper, P, proline-rich region). The first box (broken lines) shows the noncoding region of the *AT* gene. The consequences of two mutations are unknown (denoted *). The exons are not to scale due to their large number. (b) Mutations in the ATM protein. Notation as in Figure 4.1.

b.

```
          T           X                        X            X                   X   C
1         MSLVLNDLLI CCRQLEHDRA TERKKEVEKF KRLIRDPETI KHLDRHSDSK QGKYLNWDAV FRFLQKYIQK ETECLRIAKP NVSASTQASR QKEMQEISSL
                                                                                        X      #    X                                            L
                    XX              "CLX"                      X             X     XX  X                                            X
101       VKYFIKCANR RAFRLKQEL LNYIMDTVKD SSNGAIYGAD CSNILLKDIL SVRKYWCEIS QQQMLELFSV YFRLYLKPSQ DVHRVLVARI IHAVTKGCCS

                X                                                 X                                                  X
201       QTDGLNSKFL DFFSKAIQCA RQEKSSSGLN HILAALTIFL KTLAVNFRIR VCELGDEILP TLLYIWTQHR LNDSLKEVII ELFQLQIYIH HPKGAKTQEK
                                                                                    X                   X              X
          X                                                              X                       X
301       GAYESTKWRS ILYNLYDLLV NELSHIGSRG KYSSGFRNIA VKENLIELNA DICHQVFNED TRSLELSQSY TTTQRESSDY SVCKRKKIE LGWEVIKDHL
                                                                                                  S
401       QKSQNPEDLV PWLQIATQLI SKYPASLPNC ELSPLIMILS QLLPQQRHGE RTPVVLRCLT EVALCQDKRS NLESSQKSDL LKLWNKIWCI TFRGISSEQI
            X                                                                                                              X
501       QAENFGLLGA IIQGSLVEVD REEPKLFTGS ACRSCPAVC CITLALTTSI VPGAVRMGIE QNMCEVNRSF SLKESIMKWL LFYQLEGDLE NSTEVPFILH
                                                                        @
601       SNFPHLVLEK ILVSLTMKNC KAAMNFFQSV PECEHHQKDK EELSFSEVEE LFLQTTFDRM DFLTIVREGG IEKHQSSIGF SVHQNLKESL DRCLLGLSEQ
                                                                        X                    X          D              C              #
          "FIP"
          X   #
701       LLINIYSSEIT NSETLVKCSR LLVGVLGCVC YMGVLAEEEA YKSELFQKAN SLMQCAGESI TLFRNKTNEE FRIGSLRNMM QLCTRCLSNC TKKSENKIAS
          X   #                                                      X  L
801       GFFLRLLTSK LMNDIADICK SLASFIKKFF DRGEVESMED DTNGNLMEVE DQSSMNLFND YPDSSVSDAN EPGESQSTIG AINPLAEEYL SKQDLLFLDM
                                                                        X
          Q                                                                              X
901       LKFLCLCVTT AQTNTVSFRA ADIRRKLIML IDSTLEPTK SLHLHMVLML LKELPGEEYP LMEDVLELL KPLSNVCSLY RKDQDVCKTI LNHVLHVVKN
                          #    X                  X   X                X                  X
1001      LGQSNMDSEN TRDAQGQFLT VIGAFWHLTK ERKYIFSVRM ALVNCLKTLL EADPYSKWAI LNVMGKDFPV NEVFTQFLAD NHHQVRMLAA ESINRLFQDT
                            X                            #                                                                        K
          X   W                                  X                                              X   X          X              C
1101      KGDSSRLLKA LPLRLQQTAF ENAYLRAQEG YREMSHSAEN PEILDEIYNR KSVLITILAV VLSCSPICEK QALFALCKSV KENGLEPHLV KKVIEKVSET
          X  X                                                                X                          X
1201      FGYRRLEDFM ASHLDLYLVLE WLNLQDTEYN LSSFPFILLN YTNIEDEYRS CYKVLIPHLV IRSHFDEVKS IANQIQEDWK SLLTDCFPKI LVNILPYFAY
                                                              X                    E          X   X
1301      EGTRDSGMAQ QRETATKVYD MLKSENLLGK QIDHLFISNL PEIYVELLMT LHEPANSSAS QSTDLCDFSG DLDPAPNFPH FFSHVIKATF AYISNCHKTK
                                                                                                  R
          X                          #                          X   X   X
1401      LKSILEILSK SFDGYQKILL AICGQAAETN NVYKKHRILK IYHLFVSLLL KDIKSGLGA WAFVLRDVIY TLIHYINQRP SCIMDVSLRS FSLCCDLLSQ
                                                                                                                        #
          X
1501      VCQTAVTYCK DALENLHLHVI VGTLIPLVYE QYEVQKQVLD LLKYLVIDRK DNENLYITIK LLDFFPDHVV FKDLRITQQK IKTYSRGFFSL LEEINHFLSV
```

Figure 4.10. (Continued)

153

c.

```
                          X X                  #                                                           R
1601    SVYDALPLTR LEGLKDLRRQ LELHKDQMVD IMRASQDNPQ DGIMVKLVVN LLQLSKMAIN HTGEKEVLEA VGSCLGEVGP IDFSTIAIQH SKDASYTKAL
                                                                                        @
                                                                                        X
                                   X   X              I                      X     X    X X             X
1701    KLFEDKELQW TFIMLTYLNN TLVEDCVKVR SAAVTCLKNI LATKTGHSFW EIYKMTTDPM LAYLQPFRTS RKKFLEVPRF DKENPFEGLD DINLWIPLSE
        "QX"         VX                          X        X XX  X                        X         X        X
1801    NHDIWIKTLT CAFLDSGGTK CEILQLLKPM CEVKTDFCQT VLPYLIHDIL LQDTNESWRN LLSTHVQGFF TSCLRHFSQT SRSTTPANLD SESEHFFRCC
                                                                                   X                         X
            X        G                 X       X                        X          X   X    X              X X
1901    LDKKSQRTML AVVDYMRRQK RPSSGTIFND AFWLDLNYLE VAKVAQSCAA HFTALLYAEI YADKKSMDDQ EKRSLAFEEG SQSTTISSLS EKSKEETGIS
         "TTNIX"                                                               D
           XX         X G  X              X X                   K               X
2001    LQDLLLEIYR SIGEPDSLYG CGGGKMLQPI TRLRTYEHEA MWGKALVTYD LETAIPSSTR QAGIIQALQN LGLCHILSVY LKGLDYENKD WCPELEELHY
                                                                                                           X
                        #        X           XX  X                                                         #
2101    QAAWRNMQWD HCTSVSKEVE GTSYHESLYN ALQSLRDREF STFYESLKYA RVKEVEEMCK RSLESVYSLY PTLSRLQAIG ELESIGELFS RSVTHRQLSE
                                "I  S"
                                    C
                                                        "       H"        X        X                     X
2201    VYIKWQKHSQ LLKDSDFSFQ EPIMALRTVI LEILMEKEMD NSQRECIKDI LTKHLVELSI LARTFKNTQL PERAIFQIKQ YNSVSCGVSE WQLEEAQVFW
             X                     #   X   XX  X                                                      X
2301    AKKEQSLALS ILKQMIKKLD ASCAANNPSL KLTYTECLRV CGNWLAETCL ENPAVIMQTY LEKAVEVAGN YDGESSDELR NGKMKAFLSL ARFSDTQYQR
                                 G
                                 X  #                 X                            D              X
2401    IENYMKSSEF ENKQALLKRA KEEVGLLREH KIQTNRYTVK VQRELELDEL ALRALKEDRK RFLCKAVENY INCLLSGEEH DMWVFRLCSL WLENSGVSEV
             X        X                                #  #       D X             X                             E X
2501    NGMMKRDGMK IPTYKFLPLM YQLAARMGTK MMGGLGFHEV LNNLISRISM DHPHHTLFII LALANANRDE FLTKPEVARR SRITKNVPKQ SSQLDEDRTE
                               V                                                 X
                 X             X "EP"X      N        SX             P       #    G                                  X
2601    AANRIICTIR SRRPQMVRSV EALCDAYIIL ANLDATQWKT QRKGINIPAD QPITKLKNLE DVVVPTMEIK VDHTGEYGNL VTIQSFKAEF RLAGGVNLPK
                                                                      X
                      X  R#        V          X                     X #   X       X                        X   X X
2701    IIDCVGSDGK ERRQLVKGRD DLRQDAVMQQ VFQMCNTLLQ RNTETRKRKL TICTYKVVPL SQRSGVLEWC TGTVPIGEFL VNNEDGAHKR YRPNDFSAFQ
                                                            X      R
           X       X      #      Y XC     C  #              P     "RI"  #       R                    X
2801    CQKKMMEVQK KSFEEKYEVF MDVCQNFQPV FRYFCMEKFL DPAIWFEKRL AYTRSVATSS IVGYILGLGD RHVQNILINE QSAELVHIDL GVAFEQGKIL
           #                                                                                                "HSQX"
           G  X                XX      X X      X       X                                         X          X
2901    PTPETVPFRL TRDIVDGMGI TGVEGVFRRC CEKTMEVMRN SQETL LTIVE VLLYDPLFDW TMNPLKALYL QQRPEDETEL HPTLNADDQE CKRNLSDIDQ
        X  X  XC X                                        X
3001    SFDKVAERVL MRLQEKLKGV EEGTVLSVGG QVNLLIQQAI DPKNLSRLFP GWKAWV
```

Figure 4.10. (*Continued*)

moiety, which gives the protein the specific absorbance characteristics of a flavo-cytochrome (Cross *et al.*, 1995; Segal *et al.*, 1992). The gp91^{phox}/p22^{phox} complex is also called flavocytochrome b_{558} (for its α absorbance peak at 558 nm) or flavo-cytochrome b_{-245} (for its midpoint potential of -245 mV). gp91^{phox} Is an integral membrane protein; it accepts electrons from NADPH at its cytosolic, C-terminal region and it transmits these electrons via FAD and the heme groups to molec-ular oxygen at its membrane-bound, N-terminal region. In this way, superoxide (O_2^-) is generated in the phagosome that contains the ingested microorganisms. Superoxide is spontaneously converted into hydrogen peroxide (H_2O_2) and also forms other strongly microbicidal compounds.

Superoxide and its derivatives can cause severe tissue damage, and the potentially dangerous NADPH oxidase activity is therefore tightly controlled. In resting leukocytes, the flavocytochrome is present in the membranes of spe-cific granules and secretory vesicles. Upon phagocytosis of microorganisms, these organelles fuse with the phagosomal membrane surrounding the ingested mate-rial. At the same time, several cytosolic proteins (p40^{phox}, p47^{phox}, and p67^{phox}) translocate to the membrane and form a complex with the flavocytochrome (Heyworth *et al.*, 1991; Leto *et al.*, 1994; Leusen *et al.*, 1994a). This complex formation changes the structure of the flavocytochrome, thus enabling it to bind NADPH and to start its enzymatic action. The whole activation process is ini-tiated by the phosphorylation of p47^{phox}, and possibly of other *phox* components, and is regulated by several low-molecular-weight GTP-binding proteins and their nucleotide exchange factors (Abo *et al.*, 1991; Segal *et al.*, 1985). Defects in p22^{phox}, p47^{phox}, or p67^{phox} also cause CGD, but these forms of the disease are autosomally transmitted and less common than XCGD (Roos *et al.*, 1996b).

The N-terminal half of gp91^{phox} contains several hydrophobic stretches that are believed to be transmembrane domains (Dinauer *et al.*, 1987; Teahan *et al.*, 1987) (Figure 4.11). This part of the protein is also involved in the associ-ation with p22^{phox}, and it contains four histidines that act as binding sites for the two hemes (structural homology with FRE1; Finegold *et al.*, 1996). Three extra-cellular asparagines are glycosylated (Wallach and Segal, 1997). The C-terminal half of gp91^{phox} shows sequence similarity with a family of flavoproteins, which also includes spinach NADP ferredoxin reductase, especially in the FAD- and NADPH-binding regions (Segal *et al.*, 1992). The three-dimensional structure of the C-terminus of gp91^{phox} has been modeled (Taylor *et al.*, 1993).

The disease-causing mutations and mutation types are indicated in Figure 4.11. There are a total of 459 patients in the database, from 416 unre-lated families, with 286 unique mutations (69%) (Table 4.3). XCGD mutations are scattered throughout the *CYBB* gene (Figure 4.11), but there is overrepre-sentation of mutations in the N-terminal segment and underrepresentation in the NADPH-binding region, calculated on the basis of the number of unrelated families. One polymorphism has been described, affecting position 1102 (G/C)

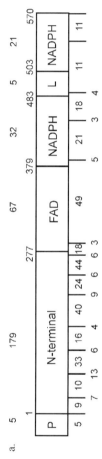

Figure 4.11. (a) Numbers of nonrelated families with mutations in gp91^{phox} arranged according to protein coding domains (top) and according to exons and introns (below). P, promoter region; FAD, FAD-binding domain; NADPH, NADPH-binding domain; L, loop-over NADPH-binding domain. (b) Location of individual mutations in gp91^{phox}. Mutation that leads to an mRNA splice defect are denote with *. Other notations as in Figure 4.1.

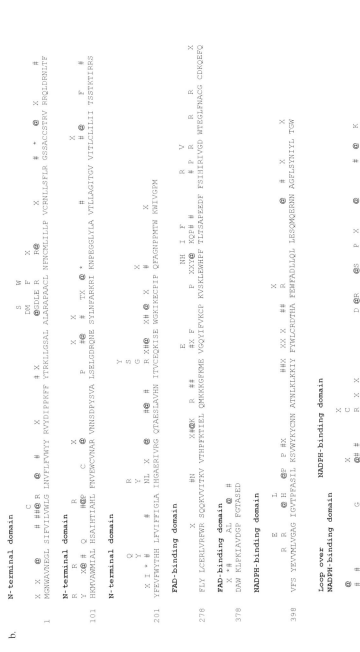

Figure 4.11. (*Continued*)

(numbering the first nucleotide in the start codon as 13). Two *Nsi*I-specific RFLPs are also present, as well as several highly polymorphic $(GT/AC)_n$ repeats, but their exact localization is not known.

Missense mutations comprise the most common mutational event, followed by nonsense mutations and deletions. Frameshift mutations constitute the largest group when deletions, insertions, and splice site mutations are added up. Nonsense mutations are frequently caused by mutations in the CpG dinucleotides, especially in the CGA codon for arginine. Substitutions in three families alter the start codon and prevent expression of the protein. The most frequently affected sites in point mutations are CpG dinucleotides. The 27 CpG dinucleotides in the coding region comprise only 3.2% of the *CYBB* gene, although CG-to-TG or -CA mutations constitute 31.5% of the single-base substitutions.

In five families, nucleotide substitutions in the promoter region of *CYBB* have been found, leading to absence of gp91^{phox} expression. These mutations affect binding sites for transcription factors, one of which has been identified as PU.1 (Suzuki *et al.*, 1998). Mutations at exon–intron boundaries in 79 families cause aberrant mRNA splicing, although this was proven in only about half of these cases (in the other half, only genomic DNA was investigated). In addition, single nucleotide changes in the middle of exon 6 and in the middle of intron 6 have created 3' (acceptor) splice sites, causing a partial exon deletion and insertion of 94 bp, respectively, in the mRNA (de Boer *et al.*, 1992; Noack *et al.*, 1999). Even in cases of in frame exon deletions, expression of curtailed gp91^{phox} protein has never been reported. Insertions and deletions have been identified in 38 and 92 families, respectively. Two large insertions, consisting of transposon sequences, were found, one of which is located in an intron and causes aberrant mRNA splicing.

The larger deletions encompass multiple exons, often the whole *CYBB* gene, and sometimes even multiple genes; the latter can cause additional clinical symptoms besides those typical for XCGD (e.g., Duchenne muscular dystrophy, X-linked retinitis pigmentosa, McLeod hemolytic anemia).

Most of the data in the CYBBbase concern severe (classical) CGD. Most of the mutations lead to absence of gp91^{phox} protein expression (X91^{o} CGD), through decreased mRNA stability and/or decreased protein stability; the latter may be due to intrinsic instability or decreased interaction with the stabilizing p22^{phox} protein. In about 10% of the cases, with small in-frame deletions or missense mutations, the expression of gp91^{phox} is strongly diminished, with concomitantly diminished NADPH-oxidase activity (X91^{-} CGD). In 17 families with missense mutations, the gp91^{phox} protein is normally expressed but totally lacks activity (X91^{+} CGD). Detailed study of some of these cases has revealed interesting information about the interaction of the protein with its prosthetic groups or with other NADPH oxidase components (Cross *et al.*, 1995; Leusen *et al.*, 1994a, 2000).

The disease is considered to have full penetrance. XCGD in females with highly skewed lyonization has been reported in four individuals with either missense (3) or nonsense (1) mutations. The severity of XCGD can be classified according to susceptibility to infections or can be based on remaining gp91^{phox} expression and NADPH-oxidase activity. In general, CGD patients with mutations in autosomal NADPH-oxidase genes present with clinical symptoms at a later age and with milder clinical problems than XCGD patients, perhaps due to slightly more remaining oxidase activity in the phagocytes from autosomal CGD patients. Most of the data in the CYBBbase is for severe (classical) CGD. However, no clear correlation between clinical symptoms and cellular defects is found when X91^{o} and X91^{-} CGD patients are compared, indicating that modifier genes may be involved. Indeed, in a survey of 129 clinically well-defined CGD patients (25 autosomal and 104 XCGD), Foster *et al.* (1998) found an increased risk for gastrointestinal complications associated with a polymorphism in the promoter region of myeloperoxidase and with the NA2 allele of FcγRIIIb (Foster *et al.*, 1998). The risk for autoimmune complications was associated with variant alleles of mannose-binding lectin.

B. Mutations in Autosomal Chronic Granulomatous Disease (CYBAbase, NCF1base, NCF2base)

Autosomal chronic granulomatous disease (A-CGD) is a hereditary immunodeficiency caused by mutations in any one of three genes coding for proteins of NADPH oxidase, the enzyme in phagocytic leukocytes that generates superoxide (Roos and Curnutte, 1999). These genes are CYBA for p22^{phox}, NCF1 for p47^{phox}, and NCF2 for p67^{phox} (*phox* stands for *phagocyte oxidase*) (Table 4.2). The p22^{phox} protein is a transmembrane protein that associates with gp91^{phox}, the enzymatic subunit of the NADPH oxidase (defects in gp91^{phox} cause X-linked CGD). p47^{phox} and p67^{phox} are cytosolic proteins that associate with the gp91^{phox}/p22^{phox} complex in phagocytes that have ingested microorganisms or have been stimulated by high concentrations of chemoattractants. This association activates the NADPH oxidase enzyme, leading to the reduction of molecular oxygen to form superoxide. For more general information on NADPH oxidase and CGD, see XCGD.

1. *CYBA* mutations

The p22^{phox} protein stabilizes gp91^{phox} during maturation and processing: a deficiency of p22^{phox} leads to concomitant lack of gp91^{phox} expression in phagocytes, and vice versa (Porter *et al.*, 1994; Yu *et al.*, 1997). The N-terminal part of p22^{phox} crosses the membrane at least three times, and the cytoplasmic C-terminus contains a proline-rich domain to which the N-terminal SH3 domain of p47^{phox} can bind (Leto *et al.*, 1994; Sumimoto *et al.*, 1994) (Figure 4.12). In activated

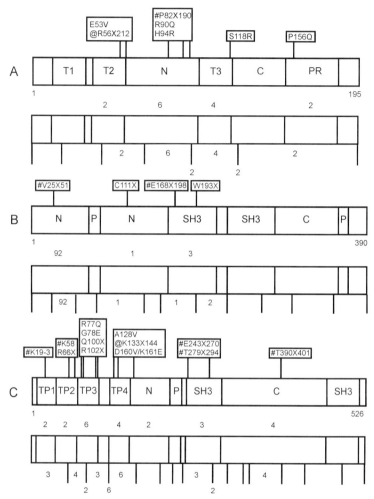

Figure 4.12. (A) Location of individual mutations in p22phox. Numbers of alleles in nonrelated families with mutations in p22phox. T, transmembrane domain; N, N-terminal domain; C, C-terminal domain; PR, proline-rich domain. (B) Location of individual mutations in p47phox. Numbers of alleles in nonrelated families with mutations in p47phox. No intronic mutations have been identified in p47phox. Details are the same as in (A) except that P, proline-rich domain. (C) Location of individual mutations in p67phox. Numbers of alleles in nonrelated families with mutations in p67phox. Details are the same as in (A). TP1–TP4 = tetratricopeptide repeats 1–4.

phagocytes, therefore, p22^{phox} appears to act as a docking protein, but p22^{phox} is also expressed in many other cells, without expression of gp91^{phox}. The function of p22phox in these other cells is unknown.

The mutations in *CYBA* that cause autosomal CGD (A22-CGD) are indicated in Figure 4.12. There are a total of 12 patients in the database, representing 10 unrelated families (20 alleles), and 10 unique mutations (50%) (Table 4.3). In 8 families, mutations are found in a homozygous state, and in 2 families the patients are compound heterozygotes. The most common type of defect is a missense mutation (12), followed by four splice-site mutations, three frameshift deletions, and one frameshift insertion. Six of the mutations are in the N-terminal domain, with the central and most C-terminal transmembrane domains containing two and four mutations, respectively. The proline-rich domain contains two mutations, but none occur in the rest of the cytoplasmic region. One mutation, which is homozygous in one patient, results in a gross deletion of the gene. This is the only case in which mRNA for p22^{phox} is undetectable; in all other cases, it is present. Nevertheless, in all but one patient the p22^{phox} protein is completely absent, and only three polymorphisms have been described that result in amino acid substitutions (K60T, H72Y, A174V). Apparently, small changes in the structure of p22^{phox} lead to its intrinsic instability or to an instability resulting from a loss of coupling to gp91^{phox}. In the one patient with full expression of aberrant p22^{phox}, the mutation (both alleles) is in the proline-rich domain, which serves as a binding site for the SH3 domain in p47^{phox}. The P156Q mutation in p22^{phox} therefore destroys the docking site for the cytosolic NADPH components, and as a result, the enzyme cannot be activated (Leusen *et al.*, 1994b).

2. NCF1 mutations

The p47^{phox} protein forms a complex in the cytosol with p67^{phox} and p40^{phox}, held together, at least in part, by interactions between SH3 and proline-rich domains, within and between the components. Activation of phagocytes (e.g., by binding of IgG-covered microorganisms to Fcγ receptors on the cell surface) initiates a number of signal transduction events, including phosphorylation of serine residues in p47^{phox}. This leads in turn to a conformational change and exposure of an SH3 domain in p47^{phox} that can interact with p22^{phox}. This process is essential for initiation of the enzymatic activity of NADPH oxidase. The N-terminus and the C-terminus of the p47^{phox} protein each contain one SH3 domain and one proline-rich region.

The mutations in *NCF1* that cause autosomal CGD (A47-CGD) are shown in Figure 4.12. There are a total of 51 patients in the database, from 48 unrelated families (96 alleles), with four unique molecular defects identified to date (4%) (Table 4.3). The vast majority of alleles (>92%) contains a deletion of a dinucleotide (GT) at the start of exon 2 that causes a frameshift and a premature stop codon (Casimir *et al.*, 1991). This mutation originates in one or more highly homologous pseudo-genes of p47^{phox}, and it is likely that recombination events between *NCF1* and its pseudo-genes are the main cause of A47-CGD (Görlach *et al.*, 1997). In those patients with the GT deletion who have been studied, the size and quantity of mRNA is grossly normal. There is one other unique mutation in the N-terminal region of the protein and two in the N-terminal SH3 domain. The most common mutations are frameshift deletions (87), followed by three nonsense mutations.

3. NCF2 mutations

The p67^{phox} protein contains two C-terminal SH3 domains, one of which is able to associate with a proline-rich domain in p47^{phox}. The p67^{phox} protein also contains four tetratricopeptide repeats (TPRs) and a proline-rich domain at the N-terminus. Both p47^{phox} and p67^{phox} associate with the p22^{phox}/gp91^{phox} complex in the membrane after phagocyte activation. The interaction of p67^{phox} with gp91^{phox} facilitates the flow of electrons from NADPH to FAD, whereas p47^{phox} may be involved in facilitating electron flow from FAD to oxygen (Cross and Curnutte, 1995; Nisimoto *et al.*, 1999).

The mutations in *NCF2* that cause autosomal CGD (A67-CGD) are shown in Figure 4.12. There are 20 patients in the database, from 17 unrelated families (34 alleles), with a total of 17 unique molecular defects (50%) (Table 4.3). Patients in 12 families are homozygous for their mutation; in the remaining 5 families they are compound heterozygotes. Deletions are the most common type of mutation in A67-CGD, with 3 in-frame and 8 resulting in a frameshift. Splice site mutations are the next most frequent mutation (10), followed by missense mutations (7), nonsense mutations (4), and frameshift insertions (2). Many mutations predict changes in the TPR-containing region of the protein, with two in TPR1, two in TPR2, six in TPR3, and four in TPR4. There are also two mutations in the remainder of the N-terminus, three in the N-terminal SH3 domain, and four in the C-terminus. With the exception of one patient, the mutant p67^{phox} protein is never detectable. The one exception is a compound heterozygote for a large deletion and a triplet nucleotide deletion that is predicted to remove K-58 from the protein (Leusen *et al.*, 1996). Presumably, the first mutation does not lead to protein expression, but the second does. K-58 is in the *Rac1/2*-binding region of p67^{phox}, and removal of this amino acid therefore destroys the proper activation of the NADPH oxidase.

C. Mutations of IFN-γ-Mediated Immunity

Bacillus Calmette-Guérin (BCG) vaccines and environmental nontuberculous mycobacteria (NTM) may cause severe disease in otherwise healthy children with no overt immunodeficiency (Casanova et al., 1995, 1996; Frucht and Holland, 1996; Levin et al., 1995). Patients with idiopathic BCG and NTM infections do not generally have associated infections, apart from salmonellosis, which affects less than half of the cases. Parental consanguinity and familial forms are frequently observed, and this syndrome was therefore designated as "Mendelian susceptibility to mycobacterial infection" (OMIM 209950). The genetic basis of the syndrome does not seem to be the same in all affected families. In most familial cases, inheritance is autosomal and recessive, but X-linked recessive inheritance seems to be involved in one family (Frucht and Holland, 1996) and autosomal dominant inheritance has been reported for two other families (Jouanguy et al., 1999). Clinical outcome differs between patients and has been found to correlate with the type of BCG granulomatous lesion present (Emile et al., 1997). Children with lepromatous-like granulomas (poorly delimited, multibacillary, with no epithelioid or giant cells) generally die of overwhelming infection, whereas children with tuberculoid granulomas (well delimited, paucibacillary, with epithelioid and giant cells) have a favorable outcome. Genetic heterogeneity was therefore suspected.

Four genes have been found to be mutated in patients with this syndrome (Table 4.2): IFNGR1 and IFNGR2, encoding the two chains of the receptor for IFNγ, a pleiotropic cytokine secreted by NK and T cells; IL12B, encoding the p40 subunit of IL-12, a potent IFNγ-inducing cytokine secreted by macrophages and dendritic cells; and IL12RB1, encoding the β1 chain of the receptor for IL-12, expressed on NK and T cells (Figure 4.13). The type of mutation may also partly account for clinical heterogeneity: dominant and recessive mutations have been found in IFNGR1, and null mutations and mutations with mild effects have been found in IFNGR1 and IFNGR2 genes. Null mutations in IFNGR1 can be further divided in two types, depending on the pathogenic mechanism. These mutations define eight disorders, which all impair the IFNγ-mediated immunity. Impaired secretion of IFNγ occurs in IL-12p40 and IL-12Rβ1 deficiency, and impaired response to IFNγ in IFNγR1 and IFNγR2 deficiency. Certain patients with the syndrome have no defect in any of the four genes identified to date.

1. IFNGR1 mutations

Complete IFNγR1 deficiency results in a selective susceptibility to early-onset and severe mycobacterial infection (Casanova et al., 1999). Lepromatous-like lesions, particularly in response to BCG, are suggestive of the absence of IFNγ-mediated immunity, whereas tuberculoid granulomas almost certainly rule out complete IFNγR1 deficiency. Bone marrow transplantation is the treatment of choice, because IFNγ treatment is ineffective. Complete IFNγR1 deficiency has been identified in nine kindreds (17 patients) (Altare et al., 1998b; Holland et al., 1998; Jouanguy et al., 1996, 2000; Newport et al., 1996; Pierre-Audigier et al., 1997; Roesler et al., 1999) (Table 4.3).

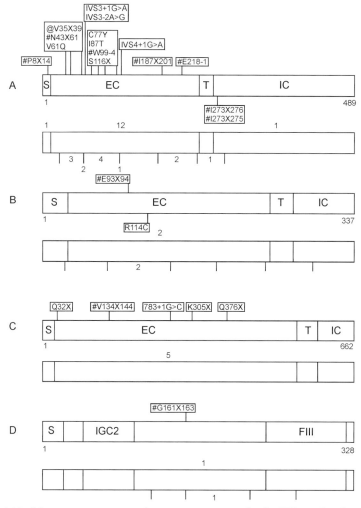

Figure 4.13. Schematic representation of mutations in proteins related to IFNγ-mediated immunity. (A) IFNGR1, (B) IFNGR2, and (C) IL12RB1 proteins contain, from the N-terminus, a signal peptide (S), an extracellular domain (EC), an transmembrane domain (T), and an intracytoplasmic domain (IC). The IL12B (D) protein contains also a signal peptide (S), IG-like C2-type domain. Numbers of alleles in nonrelated families in domains (above); in exons and at splice sites (below).

Both homozygous and compound heterozygous patients have been found, and no recurrent mutations have been observed. The mutations in 13 patients and six kindreds are null, as they preclude cell surface expression of the receptor due to a premature stop codon upstream from the segment encoding the transmembrane domain (nonsense and splice mutations, frameshift small deletions and insertions) (see Figure 4.13).

In three other families, four children with complete IFNγR1 deficiency were found to have normal expression of IFNγR1 molecules on the cell surface (Jouanguy et al., 2000). Mutations (in-frame small deletions, missense mutations in the segment encoding the extracellular ligand-binding domain) prevented the binding of the encoded surface receptors to their natural ligand, IFNγ (see Figure 4.13).

Cells from the children with partial IFNγR1 deficiency respond to IFNγ at high concentrations. The clinical phenotype is milder than that of children with complete IFNγR1 deficiency. There is a correlation between the genotype (null or mild mutation), the cellular phenotype (complete or partial defect), the histological phenotype (immature or mature granulomas), and the clinical phenotype (poor or favorable outcome) (Lamhamedi et al., 1998). Given the good prognosis, bone marrow transplantation is not indicated. As IFNγ induces signaling events in vitro, IFNγ therapy is probably the best option for patients with partial IFNγR1 deficiency who may suffer from mycobacterial disease that is refractory to antibiotics.

Partial IFNγR1 deficiency may be caused by recessive or dominant IFNGR1 alleles. Two siblings with partial recessive IFNγR1 deficiency have been reported (Jouanguy et al., 1997). A homozygous recessive missense mutation causing an aminoacid substitution in the extracellular domain of the receptor was identified (Figure 4.13). The mutation probably reduces but does not abolish the affinity of the encoded cell surface receptor for its ligand, IFNγ.

Eighteen patients from 12 unrelated kindreds were found to have a dominant form of partial IFNγR1 deficiency (Jouanguy et al., 1999) (Table 4.3). These patients have a heterozygous frameshift small deletion in IFNGR1 exon 6, downstream from the segment encoding the transmembrane domain. The mutant alleles encode truncated receptors with only five intracellular amino acids. The receptors reach the cell surface and bind IFNγ with normal affinity. The receptors dimerize and form a tetramer with two IFNγR2 molecules, but they do not transduce IFNγ-triggered signals due to the lack of intracellular binding sites for the cytosolic molecules (JAK-1 and STAT-1) involved in the signaling cascade. The receptors also accumulate at the cell surface due to the lack of an intracellular recycling site. The combination of normal binding to IFNγ, abolished signaling in response to IFNγ, and accumulation of receptors at the cell surface accounts for their dominant-negative effect. Most IFNγR1 dimers in heterozygous cells are nonfunctional due to the presence of at least one defective molecule. The few wild-type IFNγR1 dimers that form in response to IFNγ account for the defect being partial rather than complete. A most interesting genetic feature of this disorder is that position 818 of *IFNGR1* is the first small deletion hotspot to be identified in the human genome. Overlapping small deletions (#K272X275 in 11 kindreds and #I273X276 in one) were found to occur independently in 12 unrelated families. A model of slipped mispairing events and subsequent repair during replication was proposed, based on the presence of two direct repeats and small deletion consensus motifs in the vicinity of nucleotide 818.

2. *IFNGR2* mutations

One child with complete IFNγR2 deficiency has been reported (Dorman and Holland, 1998). A homozygous recessive frameshift deletion was found in the *IFNGR2* coding region, resulting in a premature stop codon upstream the transmembrane domain. The child had early-onset and severe *Mycobacterium* infections, but no mature granulomas were observed. Thus, null recessive *IFNGR2* mutations, like null recessive *IFNGR1* mutations, may be responsible for early-onset and severe mycobacterial infection with impaired granuloma formation. The recommended treatment for patients with complete IFNγR2 deficiency is the same as that for patients with complete IFNγR1 deficiency.

A 20-year-old patient with a history of BCG and *Mycobacterium abscessus* infection was found to have partial IFNγR2 deficiency due to a homozygous missense mutation (Döffinger *et al.*, 2000) (Table 4.3, Figure 4.13). Cellular responses were impaired, but not abolished, following stimulation with IFNγ. The molecular mechanism remains to be determined. This case illustrates, as for IFNγR1 deficiency, that there is a strict correlation between the *IFNGR2* genotype and the cellular, histological, and clinical phenotype. The level of IFNγ-mediated immunity seems to be the crucial factor determining the histopathological lesions associated with, and the clinical outcome of, mycobacterial infections.

3. *IL12B* mutations

A child with a mild histopathological and clinical phenotype and a recessive mutation in the *IL12B* gene has been reported (Altare *et al.*, 1998c). The mutation consists of a homozygous frameshift deletion of 4.4 kb encompassing two coding exons (Figure 4.13). Three nucleotides adjacent to the recombination breakpoints were identical and may have contributed to the recombination process. Transfection of a defective cell line with the wild-type *IL12B* gene led to the secretion of IL-12 p40 and p70. This implies that there is a causative relationship between the *IL12P40* homozygous deletion and the lack of IL-12 production. Impaired IFNγ secretion was complemented in a dose-dependent manner by exogenous recombinant IL-12, implying that IFNγ deficiency is not a primary event but a consequence of inherited IL-12 deficiency. Mycobacterial infections occur primarily because IFNγ-mediated immunity is impaired. Residual, IL-12-independent secretion of IFNγ probably accounts for the clinical phenotype being milder than that of children with complete IFNγR deficiency.

4. *IL12RB1* mutations

Mutations in the gene encoding the β1 subunit of the IL-12 receptor have been identified in seven patients (six kindreds) with a mild phenotype (Table 4.3, Figure 4.13) (Altare *et al.*, 1998a; de Jong *et al.*, 1998). In the absence of complementation experiments, a missense mutation observed in one family has not been validated and is not represented. In another family, a splice mutation was suspected because the transcript is 140 nucleotides shorter but the genomic mutation has not been identified yet (designated #V137X144). Patients in the other four families were homozygous for nonsense and splice null mutations that give a premature stop codon upstream from the transmembrane domain and preclude surface expression of the receptor. Impaired IFNγ secretion is probably responsible for mycobacterial disease in IL-12Rβ1-deficient children and residual, IL-12-independent, IFNγ-mediated immunity probably accounts for the milder clinical and histological phenotype.

Selective susceptibility to BCG or NTM has long been suspected to be a heterogeneous Mendelian disorder. Different types of mutation (allelic heterogeneity) in four genes (nonallelic heterogeneity), *IFNGR1*, *IFNGR2*, *IL12P40*, and *IL12RB1*, have been identified. Interestingly, the discovery of partial dominant IFNγR1 deficiency led to the identification of the first hotspot for small deletions in the human genome. The eight disorders resulting from these mutations are genetically different but immunologically similar, as impaired IFNγ-mediated immunity is the common pathogenic mechanism accounting for mycobacterial infection in all patients. Complete IFNγR1 and IFNγR2 deficiencies predispose patients to overwhelming infection with impaired granuloma formation in early childhood, whereas partial IFNγR1 and IFNγR2 deficiencies and complete IL-12 p40 and IL-12Rβ1 deficiencies predispose patients to curable infection with mature granulomas at various ages. An accurate molecular diagnosis is crucial to determine the optimal treatment strategy for individual patients.

D. Leukocyte Adhesion Deficiency (LAD) Type I

LAD I is an autosomal recessive disorder caused by mutations in the gene coding for the common β2 subunit of the β2 integrin (CD18) (Etzioni and Harlan, 1999). LAD I is caused by the absence or markedly decreased expression of the β2 integrin on the surface of leukocytes, leading to severe impairment in leukocytes adhesion to vascular endothelium and migration to sites of inflammation. The patients suffer from increased incidence of bacterial infections and impaired wound healing (Anderson *et al.*, 1995). Delayed separation of the umbilical cord with omphalitis are very common in infants suffering from LAD I. Treatment consists of antibiotics and bone marrow transplantation, which has excellent results.

β_2 Expression is restricted to leukocytes, but among subsets of leukocytes the distribution of the different heterodimers differs. LFA-1 (CD11a/CD18) is expressed by lymphocytes, monocytes, and neutrophils. MAC-1(CD11b/CD18) and p150.95 (CD11c/CD18) are expressed by myeloid cells and natural killer cells.

CD18 is encoded by a gene located at the tip of the long arm of chromosome 21q22.3 (Solomon *et al.*, 1988) (Table 4.2). The gene encoding CD18 has also been cloned and sequenced in mice, cattle, dogs, and many other animals (Kishimoto *et al.*, 1987a). A similar syndrome has been described in Holstein cattle.

β_2 Integrin belongs to the integrin adhesion molecule family of transmembrane cell surface proteins, which bind externally to matrix and membrane proteins and internally to cytoskeletal proteins and thereby communicate extracellular signals. The β chains have characteristic features (Figure 4.14). Tandem repeats of four cysteine-rich regions, which are thought to be essential for tertiary structure, are conserved among the various β chains. Approximately 100 amino acids at the NH$_2$-terminus are additional conserved units that are critical for maintance of the α/β heterodimer. CD18 has a relatively small cytoplasmic domain that contains regions capable of binding to various cytoskeletal elements. The extracellular domain is much larger, and the amino-terminal portion of the $\beta2$ subunit, containing a ligand-binding region, is folded into a loop that is stabilized by disulfide bonds.

The principal function of CD18 is participation in many leukocyte–endothelial adhesion interactions, mainly by binding to members of Ig gene superfamily (ICAMs). Further, CD18 is also important in other immunological processes such as cell-mediated cytotoxicity. CD18 belongs to both "outside-in" and "inside-out" signaling systems. Through the outside-in pathway, binding of

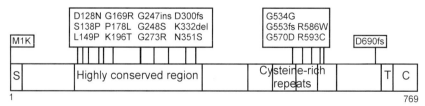

Figure 4.14. Schematic representation of mutations in LAD I. The β_2-integrin subunit contains five domains (from the N-terminus): a signal peptide (S), a highly conserved region, cysteine-rich repeats, a tranmembrane region (T), and a cytoplasmic domain (C).

extracellular-matrix proteins to β_2 integrin alters gene expression and affects cellular proliferation and differentiation. The inside-out signal transduction renders the CD18 to its active state and thus increases leukocyte adhesiveness to the endothelium (Etzioni, 1999).

The exact prevalence of LAD I is not known, and about 200 cases have been reported. In most cases no molecular analysis was done and only about 20 different mutations have been described (Figure 4.14).

The molecular basis for CD18 deficiency varies. In some cases it is due to the lack or diminished expression of CD18 mRNA. In other cases there is expression of mRNA or protein precursors of aberrant size with both larger or smaller CD18 subunits. In still other cases LAD I results a failure to process normal-sized protein precursors to the mature normal product (Wright et al., 1995). Analysis at the gene level has revealed a degree of heterogeneity which reflects this diversity. A number of point mutations have been reported, some of which lead to the biosynthesis of defective proteins with single amino acid substitutions, while others lead to splicing defects, resulting in the production of truncated and unstable proteins (Kishimoto et al., 1987a). In one kindred a 90-nucleotide deletion in the CD18 mRNA produced an in-frame deletion of a 30-amino acid region. Analysis of the genomic DNA showed this 90-bp region to be encoded by a single exon (exon 9) (Kishimoto et al., 1989).

Identified alleles span most of the coding region of the extracellular domain of CD18 and include mutations of exonic as well as intronic sequences (Table 4.3). It is significant that a disproportionate number of mutant alleles are found in the 241-amino acid region that is highly conserved among β-integrin subunits (exons 5–9) (Kishimoto et al., 1987a), or in a segment encoded by exon 13. Domains in the conserved region may represent critical contact sites between the α- and β-chain precursors. Mutations in exon 13 may impede heterodimer formation by introducing conformational changes.

In some cases homozygosity for the mutation was described, while in many others compound heterozygosity was observed. In most cases a point mutation, small insertion, or deletion in the CD18 gene have been reported. Recently, an infant with LAD I and gross abnormality in chromosome 21, representing a deletion of q22.1-3, was described (Rivera-Matos et al., 1995).

Although in most cases mutation in the CD18 gene caused defective expression of the molecule on the surface of the leukocytes, recently a mutation was described in which normal expression exists, but the molecule could not support β_2-integrin function (Hogg et al., 1999).

The clinical severity of LAD I can be divided into two types according the level of CD18 expression (Fischer et al., 1988). In the severe form, less than 1% of the normal surface expression exists and patients suffer from severe bacterial infections starting early in life, while in the moderate form 2–20% of CD18 expression in found and with milder clinical course. Mutations in the conserved

domain (exons 5–9) are associated with the severe type of phenotype involving nonconserved amino acids. Still, the spectrum of CD18 mutations provides
an incomplete explanation for the considerable heterogeneity recognized among
patients with LAD I.

VII. DISCUSSION

An analysis of PIDs reveals a spectrum of abnormalities, some of which are common to several disorders. In the 23 genes studied in this chapter, mutations have
been found from more than 2000 families (Tables 4.3 and 4.4). The numbers
of mutations vary from disease to disease. Those disorders with only a few cases
were excluded from the discussion. Diseases with X-chromosomal origin constitute about 70% of all the cases. Here a single mutated gene is enough to produce
the phenotype in males. Moreover, X-linked recessive diseases usually show full
penetrance and therefore easier diagnosis.

    All types of mutations have been identified (Tables 4.3 and 4.4). As the
numbers are increasing, the distribution within the genes is more even. However,
there are highly mutated regions, usually coding for important domains in some
disorders. As in genetic diseases in general, point mutations are the most common
alterations. Missense and nonsense mutations comprise 45% of all the genetic
defects. Truncation of the produced protein is the most common effect when
summing up nonsense, deletion, insertion, and splice site mutations.

    The proportions of mutation types are rather constant in both the X-
chromosomal and autosomal diseases except for in-frame deletions in autosomal
genes (Table 4.4). This can, however, be caused by the small sample size. The

Table 4.4. Mutation Types in Chromosomes

Chromosome	X	1	5	6	7	11	13	16	19	20	21	Autosomes total	Total
Missense	464	4	0	3	0	54	0	7	8	37	15	128	592
Nonsense	298	4	0	1	3	50	2	1	10	4	0	75	373
Deletion in frame	23	2	0	2	0	26	0	0	2	1	1	34	57
Deletion frameshift	166	3	1	5	45	123	3	1	2	4	4	191	357
Insertion in frame	9	0	0	0	0	4	0	0	0	0	1	5	14
Insertion frameshift	90	1	0	1	0	41	1	1	1	0	0	46	136
Splice site[a]	249	7	0	3	0	167	0	5	24	10	0	216	465
Gross deletions	72	1	0	0	0	1	0	1	0	0	0	3	75
Other	19	0	0	0	0	6	0	0	1	0	0	7	26
Total	1435	22	1	15	48	472	6	16	48	56	21	705	2140

[a] Total of splice site mutations.

ratios of missense and nonsense mutations are below the average and the in-frame deletions and splicing errors above the average in the autosomal PIDs.

Missense mutations appear in 28.1% of all the cases and nonsense mutations in 17.7%. Frameshift deletions are responsible for 19.3% of the genetic defects and splice site mutations 22% of the total (Table 4.4). All the other types are clearly responsible for less than 10% of mutations each. A large proportion of unique and *de novo* mutations is typical for X-linked disorders.

In several forms of PID, CpG sites are hotspots for mutations, as exemplified in the *BTK*, *IL2RG*, and *CYBB* genes. CpG dinucleotides are the single most mutated doublet, harboring, e.g., in XLA 30%, in XSCID 22%, and in XCGD 32% of the mutations despite that these dinucleotides constitute less than 5% of the gene. An increased frequency of mutations is also found in single nucleotide repeats, presumably caused by mispairing slippage as observed in the *BTK* and *IL2RG* genes. In the *IFNGR1* gene the first small deletion hotspot to be identified in the human genome was recently reported (Jouanguy *et al.*, 1999).

Mutations affecting the initiation codon are known in several PIDs, whereas alterations of the stop codon have been identified only in *SH2D1A* and *WAS*. Promoter mutations are rare in most genes, but have been identified in the *CYBB* ($n = 5$) and in the *BTK* ($n = 1$) genes. For both these promoters the binding site for the Ets transcription family member, PU.1, was affected.

Splice sites are responsible for 22% of all the mutations. Usually, the outcome is frameshift and a truncated protein. Therefore, splice site defects are only rarely detected in mild forms of PIDs. However, in certain patients sufficient proper splicing occurs resulting in mild disease.

Mutations manifest their effects in diverse ways. It is evident that, e.g., the expression of the proteins can be changed, usually lowered or prevented, and, on the other hand, stability, activity, and specificity of the proteins can be altered. Also, the production and stability of mRNA is crucial. Single amino acid substitutions may not destroy the protein fold apart from changes to function. In the future, the mRNA and protein levels, as well as enzyme activity when relevant, ought to be measured in addition to mutation detection for complete diagnosis.

It has not been possible to make genotype–phenotype correlations in many PIDs, which could be due to redundant activities at least in some disorders. Also, the number of identified cases with mild disease is in most PIDs too small for conclusive statistical analyses. Even in the diseases with substantial numbers of patients, no major conclusions have been observed except for the N-terminal part of WASP. The situation is further complicated by the varying severity of the disease even in family members bearing the same mutations. Apparently, these kinds of conclusions have to wait for larger databases as well as other information, e.g., about protein structures.

Mutation information was collected for 23 genes involved in 14 immunodeficiencies. The large number of identified mutations in unrelated families,

2140, is a remarkable undertaking for these rare disorders. During the coming years defective genes in other PIDs will also be identified. The database model presented and applied here will also be useful for the analysis of the genes to be found. Together, all these databases add considerably to our knowledge about immunodeficiencies, their genetics, diagnostics, and treatment.

Acknowledgments

Financial support from the Finnish Academy, National Technology Agency of Finland, Sigrid Juselius Foundation, Medical Resource Fund of Tampere University Hospital, Biomed 2 contract CT-983007, EU BIOTECH BIO4-CT98-0142, Telethon grants E668 and E0917, CARIPLO, Immune Deficiency Foundation, MAO/RFP N01-CP-61038 from NCI NIH (USA), NIH grant DK20902, and a grant from Enzon, Inc., Zorg Onderzoek Nederland (ZON) grant number 28-2167-2, and Swedish Medical Research Council is acknowledged. M.S.H. is a consultant to Enzon, Inc.

For reasons of lack of space it has been impossible to cite all the original work summarized here, and we therefore apologize to all those whose work could not be adequately cited.

References

Abdul-Manan, N., Aghazadeh, B., Liu, G. A., Majumdar, A., Ouerfelli, O., Siminovitch, K. A., and Rosen, M. K. (1999). Structure of Cdc42 in complex with the GTPase-binding domain of the "Wiskott-Aldrich syndrome" protein. *Nature* **399**, 379–383.

Abedi, M. R., Morgan, G., Paganelli, R., and Hammarström, L. (1995). Report from the ESID registry of primary immunodeficiencies. *In* "Progress in Immune Deficiency V" (I. Caragol, T. Español, G. Fontan, and N. Matomoros, eds.), pp. 113–115, Springer Verlag, Barcelona.

Abo, A., Pick, E., Hall, A., Totty, N., Teahan, C. G., and Segal, A. W. (1991). Activation of the NADPH oxidase involves the small GTP-binding protein p21rac1. *Nature* **353**, 668–670.

Agrawal, A., Eastman, Q. M., and Schatz, D. G. (1998). Transposition mediated by RAG1 and RAG2 and its implications for the evolution of the immune system. *Nature* **394**, 744–751.

Allen, R. C., Armitage, R. J., Conley, M. E., Rosenblatt, H., Jenkins, N. A., Copeland, N. G., Bedell, M. A., Edelhoff, S., Disteche, C. M., Simoneaux, D. K., Fanslow, W. C., Belmont, J., and Spriggs, M. K. (1993). CD40 ligand gene defects responsible for X-linked hyper-IgM syndrome. *Science* **259**, 990–993.

Altare, F., Durandy, A., Lammas, D., Emile, J. F., Lamhamedi, S., Le Deist, F., Drysdale, P., Jouanguy, E., Döffinger, R., Bernaudin, F., Jeppsson, O., Gollob, J. A., Meinl, E., Segal, A. W., Fischer, A., Kumararatne, D., and Casanova, J. L. (1998a). Impairment of mycobacterial immunity in human interleukin-12 receptor deficiency. *Science* **280**, 1432–1435.

Altare, F., Jouanguy, E., Lamhamedi-Cherradi, S., Fondanéche, M. C., Fizame, C., Ribiérre, F., Merlin, G., Dembic, Z., Schreiber, R., Lisowska-Grospierre, B., Fischer, A., Seboun, E., and Casanova, J. L. (1998b). A causative relationship between mutant IFNgR1 alleles and impaired cellular response to IFNgamma in a compound heterozygous child. *Am. J. Hum. Genet.* **62**, 723–726.

Altare, F., Lammas, D., Revy, P., Jouanguy, E., Döffinger, R., Lamhamedi, S., Drysdale, P., Scheel-Toellner, D., Girdlestone, J., Darbyshire, P., Wadhwa, M., Dockrell, H., Salmon, M., Fischer, A., Durandy, A., Casanova, J. L., and Kumararatne, D. S. (1998c). Inherited interleukin 12 deficiency in a child with bacille Calmette-Guérin and Salmonella enteritidis disseminated infection. *J. Clin. Invest.* **102**, 2035–2040.

Anderson, D. C., Kishimoto, T. K., and Smith, C. W. (1995). Leukocyte adhesion deficiency and other disorders of leukocyte adhesence and motility. In "The Metabolic and Molecular Basis of Inherited Diseases" (C. R. Scriver, A. L. Beauclet, S. Sly, and D. Volle, eds.), pp. 3955–3995. McGraw-Hill, New York.

Anonymous (1997). Primary immunodeficiency diseases. Report of a WHO scientific group. Clin. Exp. Immunol. 109, 1–28.

Antonarakis, S. E., and the Nomenclature Working Group (1998). Recommendations for a nomenclature system for human gene mutations. Hum. Mutat. 11, 1–3.

Arredondo-Vega, F. X., Santisteban, I., Daniels, S., Toutain, S., and Hershfield, M. S. (1998). Adenosine deaminase deficiency: Genotype-phenotype correlations based on expressed activity of 29 mutant alleles. Am. J. Hum. Genet. 63, 1049–1059.

Arredondo-Vega, F. X., Santisteban, I., Kelly, S., Schlossman, C. M., Umetsu, D. T., and Hershfield, M. S. (1994). Correct splicing despite mutation of the invariant first nucleotide of a 5' splice site: A possible basis for disparate clinical phenotypes in siblings with adenosine deaminase deficiency. Am. J. Hum. Genet. 54, 820–830.

Aruffo, A., Farrington, M., Hollenbaugh, D., Li, X., Milatovich, A., Nonoyama, S., Bajorath, J., Grosmaire, L. S., Stenkamp, R., Neubauer, M., Roberts, R. L., Noelle, R. J., Ledbetter, J. A., Francke, U., and Ochs, H. D. (1993). The CD40 ligand, gp39, is defective in activated T cells from patients with X-linked hyper-IgM syndrome. Cell 72, 291–300.

Bajorath, J., Chalupny, N. J., Marken, J. S., Siadak, A. W., Skonier, J., Gordon, M., Hollenbaugh, D., Noelle, R. J., Ochs, H. D., and Aruffo, A. (1995a). Identification of residues on CD40 and its ligand which are critical for the receptor-ligand interaction. Biochemistry 34, 1833–1844.

Bajorath, J., Marken, J. S., Chalupny, N. J., Spoon, T. L., Siadak, A. W., Gordon, M., Noelle, R. J., Hollenbaugh, D., and Aruffo, A. (1995b). Analysis of gp39/CD40 interactions using molecular models and site-directed mutagenesis. Biochemistry 34, 9884–9892.

Besmer, E., Mansilla-Soto, J., Cassard, S., Sawchuk, D. J., Brown, G., Sadofsky, M., Lewis, S. M., Nussenzweig, M. C., and Cortes, P. (1998). Hairpin coding end opening is mediated by RAG1 and RAG2 proteins. Mol. Cell 2, 817–828.

Bi, E., and Zigmond, S. H. (1999). Actin polymerization: Where the WASP stings. Curr. Biol. 9, R160–R63.

Blaese, R. M., Culver, K. W., Miller, A. D., Carter, C. S., Fleisher, T., Clerici, M., Shearer, G., Chang, L., Chiang, Y., Tolstoshev, P., Greenblatt J. J., Rosenberg, S. A., Klein, H., Berger, M., Mullen, C. A., Ramsey, W. J., Muul, L., Morgan, R. A., and Anderson, W. F. (1995). T lymphocyte-directed gene therapy for ADA−SCID: Initial trial results after 4 years. Science 270, 475–480.

Bontron, S., Steimle, V., Ucla, C., Eibl, M. M., and Mach, B. (1997). Two novel mutations in the MHC class II transactivator CIITA in a second patient from MHC class II deficiency complementation group A. Hum. Genet. 99, 541–546.

Bordignon, C., Notarangelo, L. D., Nobili, N., Ferrari, G., Casorati, G., Panina, P., Mazzolari, E., Maggioni, D., Rossi, C., Servida, P., Ugazio, A. G., and Mavilio, F. (1995). Gene therapy in peripheral blood lymphocytes and bone marrow for ADA-immunodeficient patients. Science 270, 470–475.

Boss, J. M. (1999). A common set of factors control the expression of the MHC class II, invariant chain, and HLA-DM genes. Microbes Infect. 1, 847–853.

Bozzi, F., Lefranc, G., Villa, A., Badolato, R., Schumacher, R. F., Khalil, G., Loiselet, J., Bresciani, S., O'Shea, J. J., Vezzoni, P., Notarangelo, L. D., and Candotti, F. (1998). Molecular and biochemical characterization of JAK3 deficiency in a patient with severe combined immunodeficiency over 20 years after bone marrow transplantation: Implications for treatment. Br. J. Haematol. 102, 1363–1366.

Brandau, O., Schuster, V., Weiss, M., Hellebrand, H., Fink, F. M., Kreczy, A., Friedrich, W.,

Strahm, B., Niemeyer, C., Belohradsky, B. H., and Meindl, A. (1999). Epstein-Barr virus-negative boys with non-Hodgkin lymphoma are mutated in the SH2D1A gene, as are patients with X-linked lymphoproliferative disease (XLP). *Hum. Mol. Genet.* **8**, 2407–2413.

Bravo, J., Staunton, D., Heath, J. K., and Jones, E. Y. (1998). Crystal structure of a cytokine-binding region of gp130. *EMBO J.* **17**, 1665–1674.

Brugnoni, D., Airò, P., Graf, D., Marconi, M., Lebowitz, M., Plebani, A., Giliani, S., Malacarne, F., Cattaneo, R., Ugazio, A. G., Albertini, A., Kroczek, R. A., and Notarangelo, L. D. (1994). Ineffective expression of CD40 ligand on cord blood T cells may contribute to poor immunoglobulin production in the newborn. *Eur. J. Immunol.* **24**, 1919–1924.

Buckley, R. H., Schiff, R. I., Schiff, S. E., Markert, M. L., Williams, L. W., Harville, T. O., Roberts, J. L., and Puck, J. M. (1997). Human severe combined immunodeficiency: Genetic, phenotypic, and functional diversity in one hundred eight infants. *J. Pediatr.* **130**, 378–387.

Buckley, R. H., Schiff, S. E., Schiff, R. I., Markert, L., Williams, L. W., Roberts, J. L., Myers, L. A., and Ward, F. E. (1999). Hematopoietic stem-cell transplantation for the treatment of severe combined immunodeficiency. *N. Engl. J. Med.* **340**, 508–516.

Callebaut, I., and Mornon, J. P. (1998). The V(D)J recombination activating protein RAG2 consists of a six-bladed propeller and a PHD fingerlike domain, as revealed by sequence analysis. *Cell. Mol. Life Sci.* **54**, 880–891.

Candotti, F., Oakes, S. A., Johnston, J. A., Giliani, S., Schumacher, R. F., Mella, P., Fiorini, M., Ugazio, A. G., Badolato, R., Notarangelo, L. D., Bozzi, F., Macchi, P., Strina, D., Vezzoni, P., Blaese, R. M., O'Shea, J. J., and Villa, A. (1997). Structural and functional basis for JAK3-deficient severe combined immunodeficiency. *Blood* **90**, 3996–4003.

Cao, X., Kozak, C. A., Liu, Y. J., Noguchi, M., O'Connell, E., and Leonard, W. J. (1993). Characterization of cDNAs encoding the murine interleukin 2 receptor (IL-2R) gamma chain: Chromosomal mapping and tissue specificity of IL-2R gamma chain expression. *Proc. Natl. Acad. Sci. (USA)* **90**, 8464–8468.

Casanova, J. L., Blanche, S., Emile, J. F., Jouanguy, E., Lamhamedi, S., Altare, F., Stéphan, J. L., Bernaudin, F., Bordigoni, P., Turck, D., Lachaux, A., Albertini, M., Bourrillon, A., Dommergues, J. P., Pocidalo, M. A., Le Deist, F., Gaillard, J. L., Griscelli, C., and Fischer, A. (1996). Idiopathic disseminated bacillus Calmette-Guérin infection: A French national retrospective study. *Pediatrics* **98**, 774–778.

Casanova, J. L., Jouanguy, E., Lamhamedi, S., Blanche, S., and Fischer, A. (1995). Immunological conditions of children with BCG disseminated infection. *Lancet* **346**, 581.

Casanova, J. L., Newport, M., Fischer, A., and Levin, M. (1999). Inherited interferon gamma receptor deficiency. *In* "Primary Immunodeficiency Diseases. A Molecular and Genetic Approach" (H. D. Ochs, C. I. E. Smith, and J. M. Puck, eds.), pp. 209–221. Oxford University Press, New York, Oxford.

Casimir, C. M., Bu-Ghanim, H. N., Rodaway, A. R., Bentley, D. L., Rowe, P., and Segal, A. W. (1991). Autosomal recessive chronic granulomatous disease caused by deletion at a dinucleotide repeat. *Proc. Natl. Acad. Sci. (USA)* **88**, 2753–2757.

Castellví-Bel, S., Sheikhavandi, S., Telatar, M., Tai, L. Q., Hwang, M., Wang, Z., Yang, Z., Cheng, R., and Gatti, R. A. (1999). New mutations, polymorphisms, and rare variants in the ATM gene detected by a novel SSCP strategy. *Hum. Mutat.* **14**, 156–162.

Cavazzana-Calvo, M., Hacein-Bey, S., de Saint Basile, G., Gross, F., Yvon, E., Nusbaum, P., Selz, F., Hue, C., Certain, S., Casanova, J. L., Bousso, P., Deist, F. L., and Fischer A. (2000). Gene therapy of human severe combined immunodeficiency (SCID)-X1 disease. *Science* **288**, 669–672.

Chang, C. H., Guerder, S., Hong, S. C., van Ewijk, W., and Flavell, R. A. (1996). Mice lacking the MHC class II transactivator (CIITA) show tissue-specific impairment of MHC class II expression. *Immunity* **4**, 167–178.

Clausen, B. E., Waldburger, J. M., Schwenk, F., Barras, E., Mach, B., Rajewsky, K., Förster, I., and

Reith, W. (1998). Residual MHC class II expression on mature dendritic cells and activated B cells in RFX5-deficient mice. *Immunity* **8**, 143–155.

Cocks, B. G., Chang, C. C., Carballido, J. M., Yssel, H., de Vries, J. E., and Aversa, G. (1995). A novel receptor involved in T-cell activation. *Nature* **376**, 260–263.

Coffey, A. J., Brooksbank, R. A., Brandau, O., Oohashi, T., Howell, G. R., Bye, J. M., Cahn, A. P., Durham, J., Heath, P., Wray, P., Pavitt, R., Wilkinson, J., Leversha, M., Huckle, E., Shaw-Smith, C. J., Dunham, A., Rhodes, S., Schuster, V., Porta, G., Yin, L., Serafini, P., Sylla, B., Zollo, M., Franco, B., Bolino, A., Seri, A., Lanyi, A., Davis, J. R., Webster, D., Harris, A., Lenoir, G., de St Basile, G., Jones, A., Behloradsky, B. H., Achhatz, H., Murgen, J., Fassler, R., Sumegi, J., Romeo, G., Vaudin, M., Ross, M. T., Meindl, A., and Bentley, D. R. (1998). Host response to EBV infection in X-linked lymphoproliferative disease results from mutations in an SH2-domain encoding gene. *Nature Genet.* **20**, 129–135.

Concannon, P., and Gatti, R. A. (1997). Diversity of ATM gene mutations detected in patients with ataxia-telangiectasia. *Hum Mutat.* **10**, 100–107.

Conley, M. E., Mathias, D., Treadaway, J., Minegishi, Y., and Rohrer, J. (1998). Mutations in btk in patients with presumed X-linked agammaglobulinemia. *Am. J. Hum. Genet.* **62**, 1034–1043.

Cooper, D. N., and Krawczak, M. (1993). "Human Gene Mutation." Bios Scientific, Oxford, UK.

Cross, A. R., and Curnutte, J. T. (1995). The cytosolic activating factors p47phox and p67phox have distinct roles in the regulation of electron flow in NADPH oxidase. *J. Biol. Chem.* **270**, 6543–6548.

Cross, A. R., Rae, J., and Curnutte, J. T. (1995). Cytochrome b-245 of the neutrophil superoxide-generating system contains two nonidentical hemes. Potentiometric studies of a mutant form of gp91phox. *J. Biol. Chem.* **270**, 17075–17077.

Daddona, P. E., Shewach, D. S., Kelley, W. N., Argos, P., Markham, A. F., and Orkin, S. H. (1984). Human adenosine deaminase. cDNA and complete primary amino acid sequence. *J. Biol. Chem.* **259**, 12101–12106.

Danial, N. N., Pernis, A., and Rothman, P. B. (1995). Jak-STAT signaling induced by the v-abl oncogene. *Science* **269**, 1875–1877.

de Boer, M., Bolscher, B. G., Dinauer, M. C., Orkin, S. H., Smith, C. I. E., Åhlin, A., Weening, R. S., and Roos, D. (1992). Splice site mutations are a common cause of X-linked chronic granulomatous disease. *Blood* **80**, 1553–1558.

de Jong, R., Altare, F., Haagen, I. A., Elferink, D. G., Boer, T., van Breda Vriesman, P. J., Kabel, P. J., Draaisma, J. M., van Dissel, J. T., Kroon, F. P., Casanova, J. L., and Ottenhoff, T. H. (1998). Severe mycobacterial and Salmonella infections in interleukin-12 receptor-deficient patients. *Science* **280**, 1435–1438.

de Saint Basile, G., Tabone, M. D., Durandy, A., Phan, F., Fischer, A., and Le Deist, F. (1999). CD40 ligand expression deficiency in a female carrier of the X-linked hyper-IgM syndrome as a result of X chromosome lyonization. *Eur. J. Immunol.* **29**, 367–373.

Derry, J. M., Ochs, H. D., and Francke, U. (1994). Isolation of a novel gene mutated in Wiskott-Aldrich syndrome. *Cell* **78**, 635–644.

Derry, J. M., Wiedemann, P., Blair, P., Wang, Y., Kerns, J. A., Lemahieu, V., Godfrey, V. L., Wilkinson, J. E., and Francke, U. (1995). The mouse homolog of the Wiskott-Aldrich syndrome protein (WASP) gene is highly conserved and maps near the scurfy (*sf*) mutation on the X chromosome. *Genomics* **29**, 471–477.

DeSandro, A., Nagarajan, U. M., and Boss, J. M. (1999). The bare lymphocyte syndrome: Molecular clues to the transcriptional regulation of major histocompatibility complex class II genes. *Am. J. Hum. Genet.* **65**, 279–286.

Dinauer, M. C., Orkin, S. H., Brown, R., Jesaitis, A. J., and Parkos, C. A. (1987). The glycoprotein encoded by the X-linked chronic granulomatous disease locus is a component of the neutrophil cytochrome b complex. *Nature* **327**, 717–720.

DiSanto, J. P., Bonnefoy, J. Y., Gauchat, J. F., Fischer, A., and de Saint Basile, G. (1993). CD40 ligand mutations in X-linked immunodeficiency with hyper-IgM. *Nature* **361,** 541–543.

DiSanto, J. P., Rieux-Laucat, F., Dautry-Varsat, A., Fischer, A., and de Saint Basile, G. (1994). Defective human interleukin 2 receptor gamma chain in an atypical X chromosome-linked severe combined immunodeficiency with peripheral T cells. *Proc. Natl. Acad. Sci. (USA)* **91,** 9466–9470.

Döffinger, R., Jouanguy, E., Dupuis, S., Fondanèche, M. C., Stéphan, J. L., Emile, J. F., Lamhamedi-Cherradi, S., Altare, F., Pallier, A., Barcenas-Morales, G., Meinl, E., Krause, C., Pestka, S., Schreiber, R. D., Novelli, F., and Casanova, J. L. (2000). Partial interferon-gamma receptor signaling chain deficiency in a patient with bacille Calmette-Gutérin and *Mycobacterium abscessus* infection. *J. Infect. Dis.* **181,** 379–384.

Dorman, S. E., and Holland, S. M. (1998). Mutation in the signal-transducing chain of the interferon-gamma receptor and susceptibility to mycobacterial infection. *J. Clin. Invest.* **101,** 2364–2369.

Durand, B., Sperisen, P., Emery, P., Barras, E., Zufferey, M., Mach, B., and Reith, W. (1997). RFXAP, a novel subunit of the RFX DNA binding complex is mutated in MHC class II deficiency. *EMBO J.* **16,** 1045–1055.

Easton, D. F. (1994). Cancer risks in A-T heterozygotes. *Int. J. Radiat. Biol.* **66,** S177–S182.

Emery, P., Durand, B., Mach, B., and Reith, W. (1996). RFX proteins, a novel family of DNA binding proteins conserved in the eukaryotic kingdom. *Nucleic Acids Res.* **24,** 803–807.

Emile, J. F., Patey, N., Altare, F., Lamhamedi, S., Jouanguy, E., Boman, F., Quillard, J., Lecomte-Houcke, M., Verola, O., Mousnier, J. F., Dijoud, F., Blanche, S., Fischer, A., Brousse, N., and Casanova, J. L. (1997). Correlation of granuloma structure with clinical outcome defines two types of idiopathic disseminated BCG infection. *J. Pathol.* **181,** 25–30.

Etzioni, A. (1999). Integrins—The glue of life. *Lancet* **353,** 341–343.

Etzioni, A., and Harlan, J. M. (1999). Leukocyte adhesion deficiency syndromes. *In* "Primary Immunodeficiency Diseases. A Molecular and Genetic Approach" (H. D. Ochs, C. I. E. Smith, and J. M. Puck, eds.), pp. 375–388. Oxford University Press, New York, Oxford.

Finegold, A. A., Shatwell, K. P., Segal, A. W., Klausner, R. D., and Dancis, A. (1996). Intramembrane bis-heme motif for transmembrane electron transport conserved in a yeast iron reductase and the human NADPH oxidase. *J. Biol. Chem.* **271,** 31021–31024.

Fischer, A., Landais, P., Friedrich, W., Morgan, G., Gerritsen, B., Fasth, A., Porta, F., Griscelli, C., Goldman, S. F., Levinsky, R., and Vossen, J. (1990). European experience of bone-marrow transplantation for severe combined immunodeficiency. *Lancet* **336,** 850–854.

Fischer, A., Lisowska-Grospierre, B., Anderson, D. C., and Springer, T. A. (1988). Leukocyte adhesion deficiency: Molecular basis and functional consequences. *Immunodefic. Rev.* **1,** 39–54.

Fondaneche, M. C., Villard, J., Wiszniewski, W., Jouanguy, E., Etzioni, A., Le Deist, F., Peijnenburg, A., Casanova, J. L., Reith, W., Mach, B., Fischer, A., and Lisowska-Grospierre, B. (1998). Genetic and molecular definition of complementation group D in MHC class II deficiency. *Hum. Mol. Genet.* **7,** 879–885.

Fontes, J. D., Kanazawa, S., Nekrep, N., and Peterlin, B. M. (1999). The class II transactivator CIITA is a transcriptional integrator. *Microbes Infect.* **1,** 863–869.

Foster, C. B., Lehrnbecher, T., Mol, F., Steinberg, S. M., Venzon, D. J., Walsh, T. J., Noack, D., Rae, J., Winkelstein, J. A., Curnutte, J. T., and Chanock, S. J. (1998). Host defense molecule polymorphisms influence the risk for immune-mediated complications in chronic granulomatous disease. *J. Clin. Invest.* **102,** 2146–2155.

Frucht, D. M., and Holland, S. M. (1996). Defective monocyte costimulation for IFN-gamma production in familial disseminated *Mycobacterium avium* complex infection: Abnormal IL-12 regulation. *J. Immunol.* **157,** 411–416.

Fuleihan, R., Ramesh, N., Horner, A., Ahern, D., Belshaw, P. J., Alberg, D. G., Stamenkovic, I., Harmon, W., and Geha, R. S. (1994). Cyclosporin A inhibits CD40 ligand expression in T lymphocytes. *J. Clin. Invest.* **93,** 1315–1320.

Fuleihan, R., Ramesh, N., Loh, R., Jabara, H., Rosen, R. S., Chatila, T., Fu, S. M., Stamenkovic, I., and Geha, R. S. (1993). Defective expression of the CD40 ligand in X chromosome-linked immunoglobulin deficiency with normal or elevated IgM. *Proc. Natl. Acad. Sci. (USA)* **90,** 2170–2173.

Futatani, T., Miyawaki, T., Tsukada, S., Hashimoto, S., Kunikata, T., Arai, S., Kurimoto, M., Niida, Y., Matsuoka, H., Sakiyama, Y., Iwata, T., Tsuchiya, S., Tatsuzawa, O., Yoshizaki, K., and Kishimoto, T. (1998). Deficient expression of Bruton's tyrosine kinase in monocytes from X-linked agammaglobulinemia as evaluated by a flow cytometric analysis and its clinical application to carrier detection. *Blood* **91,** 595–602.

Gajiwala, K. S., Chen, H., Cornille, F., Roques, B. P., Reith, W., Mach, B., and Burley, S. K. (2000). Structure of the winged-helix protein hRFX1 reveals a new mode of DNA binding. *Nature* **403,** 916–921.

Gilad, S., Chessa, L., Khosravi, R., Russell, P., Galanty, Y., Piane, M., Gatti, R. A., Jorgensen, T. J., Shiloh, Y., and Bar-Shira, A. (1998). Genotype-phenotype relationships in ataxia-telangiectasia and variants. *Am. J. Hum. Genet.* **62,** 551–561.

Giri, J. G., Ahdieh, M., Eisenman, J., Shanebeck, K., Grabstein, K., Kumaki, S., Namen, A., Park, L. S., Cosman, D., and Anderson, D. (1994). Utilization of the beta and gamma chains of the IL-2 receptor by the novel cytokine IL-15. *EMBO J.* **13,** 2822–2830.

Görlach, A., Lee, P. L., Roesler, J., Hopkins, P. J., Christensen, B., Green, E. D., Chanock, S. J., and Curnutte, J. T. (1997). A p47-phox pseudogene carries the most common mutation causing p47-phox-deficient chronic granulomatous disease. *J. Clin. Invest.* **100,** 1907–1918.

Grewal, I. S., and Flavell, R. A. (1996). The role of CD40 ligand in costimulation and T-cell activation. *Immunol. Rev.* **153,** 85–106.

Gurniak, C. B., and Berg, L. J. (1996). Murine JAK3 is preferentially expressed in hematopoietic tissues and lymphocyte precursor cells. *Blood* **87,** 3151–3160.

Hage, T., Sebald, W., and Reinemer, P. (1999). Crystal structure of the interleukin-4/receptor alpha chain complex reveals a mosaic binding interface. *Cell* **97,** 271–281.

Hagemann, T. L., and Kwan, S. P. (1999). The identification and characterization of two promoters and the complete genomic sequence for the Wiskott-Aldrich syndrome gene. *Biochem. Biophys. Res. Commun.* **256,** 104–109.

Hansen, R. S., Wijmenga, C., Luo, P., Stanek, A. M., Canfield, T. K., Weemaes, C. M., and Gartler, S. M. (1999). The DNMT3B DNA methyltransferase gene is mutated in the ICF immunodeficiency syndrome. *Proc. Natl. Acad. Sci. (USA)* **96,** 14412–14417.

Hansson, H., Mattsson, P. T., Allard, P., Haapaniemi, P., Vihinen, M., Smith, C. I. E., and Härd, T. (1998). Solution structure of the SH3 domain from Bruton's tyrosine kinase. *Biochemistry* **37,** 2912–2924.

Harton, J. A., Cressman, D. E., Chin, K. C., Der, C. J., and Ting, J. P. (1999). GTP binding by class II transactivator: Role in nuclear import. *Science* **285,** 1402–1405.

Henthorn, P. S., Somberg, R. L., Fimiani, V. M., Puck, J. M., Patterson, D. F., and Felsburg, P. J. (1994). IL-2R gamma gene microdeletion demonstrates that canine X-linked severe combined immunodeficiency is a homologue of the human disease. *Genomics* **23,** 69–74.

Hentze, M. W., and Kulozik, A. E. (1999). A perfect message: RNA surveillance and nonsense-mediated decay. *Cell* **96,** 307–310.

Hershfield, M. S. (1998). Adenosine deaminase deficiency: Clinical expression, molecular basis, and therapy. *Semin. Hematol.* **35,** 291–298.

Hershfield, M. S., and Mitchell, B. S. (1995). Immunodeficiency diseases caused by adenosine deaminase deficiency and purine nucleoside phosphorylase deficiency. *In* "The Metabolic and Molecular Bases of Inherited Disease" (C. R. Scriver, A. L. Beaudet, W. S. Sly, and D. Valle, eds.), pp. 1725–1768. McGraw-Hill, New York.

Heyworth, P. G., Curnutte, J. T., Nauseef, W. M., Volpp, B. D., Pearson, D. W., Rosen, H., and

Clark, R. A. (1991). Neutrophil nicotinamide adenine dinucleotide phosphate oxidase assembly. Translocation of p47-phox and p67-phox requires interaction between p47-phox and cytochrome b558. *J. Clin. Invest.* **87**, 352–356.

Higgs, H. N., Blanchoin, L., and Pollard, T. D. (1999). Influence of the C terminus of Wiskott-Aldrich syndrome protein (WASp) and the Arp2/3 complex on actin polymerization. *Biochemistry* **38**, 15212–15222.

Hiom, K., Melek, M., and Gellert, M. (1998). DNA transposition by the RAG1 and RAG2 proteins: A possible source of oncogenic translocations. *Cell* **94**, 463–470.

Hirschhorn, R., Tzall, S., and Ellenbogen, A. (1990). Hot spot mutations in adenosine deaminase deficiency. *Proc. Natl. Acad. Sci. (USA)* **87**, 6171–6175.

Hirschhorn, R., Yang, D. R., Israni, A., Huie, M. L., and Ownby, D. R. (1994). Somatic mosaicism for a newly identified splice-site mutation in a patient with adenosine deaminase-deficient immunodeficiency and spontaneous clinical recovery. *Am. J. Hum. Genet.* **55**, 59–68.

Hirschhorn, R., Yang, D. R., Puck, J. M., Huie, M. L., Jiang, C. K., and Kurlandsky, L. E. (1996). Spontaneous in vivo reversion to normal of an inherited mutation in a patient with adenosine deaminase deficiency. *Nature Genet.* **13**, 290–295.

Hoffman, S. M., Lai, K. S., Tomfohrde, J., Bowcock, A., Gordon, L. A., and Mohrenweiser, H. W. (1997). JAK3 maps to human chromosome 19p12 within a cluster of proto-oncogenes and transcription factors. *Genomics* **43**, 109–111.

Hogg, N., Stewart, M. P., Scarth, S. L., Newton, R., Shaw, J. M., Law, S. K., and Klein, N. (1999). A novel leukocyte adhesion deficiency caused by expressed but nonfunctional beta2 integrins Mac-1 and LFA-1. *J. Clin. Invest.* **103**, 97–106.

Holinski-Feder, E., Weiss, M., Brandau, O., Jedele, E. B., Nore, B., Bäckesjö, C. M., Vihinen, M., Hubbard, S. R., Belohradsky, B. H., Smith, C. I. E., and Meindl, A. (1998). Mutation screening of the BTK gene in 56 families with X-linked agammaglobulinemia (XLA): 47 unique mutations without correlation to clinical course. *Pediatrics* **101**, 276–284.

Holland, S. M., Dorman, S. E., Kwon, A., Pitha-Rowe, I. F., Frucht, D. M., Gerstberger, S. M., Noel, G. J., Vesterhus, P., Brown, M. R., and Fleisher, T. A. (1998). Abnormal regulation of interferon-gamma, interleukin-12, and tumor necrosis factor-alpha in human interferon-gamma receptor 1 deficiency. *J. Infect. Dis.* **178**, 1095–1104.

Hyvönen, M., and Saraste, M. (1997). Structure of the PH domain and Btk motif from Bruton's tyrosine kinase: Molecular explanations for X-linked agammaglobulinaemia. *EMBO J.* **16**, 3396–3404.

Jouanguy, E., Altare, F., Lamhamedi, S., Revy, P., Emile, J. F., Newport, M., Levin, M., Blanche, S., Seboun, E., Fischer, A., and Casanova, J. L. (1996). Interferon-gamma-receptor deficiency in an infant with fatal bacille Calmette-Guérin infection. *N. Engl. J. Med.* **335**, 1956–1961.

Jouanguy, E., Lamhamedi-Cherradi, S., Altare, F., Fondanèche, M. C., Tuerlinckx, D., Blanche, S., Emile, J. F., Gaillard, J. L., Schreiber, R., Levin, M., Fischer, A., Hivroz, C., and Casanova, J. L. (1997). Partial interferon-gamma receptor 1 deficiency in a child with tuberculoid bacillus Calmette-Guérin infection and a sibling with clinical tuberculosis. *J. Clin. Invest.* **100**, 2658–2664.

Jouanguy, E., Lamhamedi-Cherradi, S., Lammas, D., Dorman, S. E., Fondanèche, M. C., Dupuis, S., Döffinger, R., Altare, F., Girdlestone, J., Emile, J. F., Ducoulombier, H., Edgar, D., Clarke, J., Oxelius, V. A., Brai, M., Novelli, V., Heyne, K., Fischer, A., Holland, S. M., Kumararatne, D. S., Schreiber, R. D., and Casanova, J. L. (1999). A human IFNGR1 small deletion hotspot associated with dominant susceptibility to mycobacterial infection. *Nature Genet.* **21**, 370–378.

Jouanguy, E., Pallier, A., Dupuis, S., Döffinger, R., Fondanèche, M. C., Lamhamedi-Cherradi, S., Altare, F., Emile, J. F., Lutz, P., Landman-Parker, J., Donnadieu, J., Camcioglu, Y., and Casanova, J. L. (2000). In a novel form of IFN-γ receptor 1 deficiency, cell surface receptors fail to bind IFN-γ. *J. Clin. Invest.*, **105**, 1429–1436.

Karpusas, M., Hsu, Y. M., Wang, J. H., Thompson, J., Lederman, S., Chess, L., and Thomas, D.

(1995). 2 A crystal structure of an extracellular fragment of human CD40 ligand. *Structure* **3**, 1031–1039.

Kawamura, M., McVicar, D. W., Johnston, J. A., Blake, T. B., Chen, Y. Q., Lal, B. K., Lloyd, A. R., Kelvin, D. J., Staples, J. E., Ortaldo, J. R., and O'Shea, J. J. (1994). Molecular cloning of L-JAK, a Janus family protein-tyrosine kinase expressed in natural killer cells and activated leukocytes. *Proc. Natl. Acad. Sci. (USA)* **91**, 6374–6378.

Khanna, K. K., Keating, K. E., Kozlov, S., Scott, S., Gatei, M., Hobson, K., Taya, Y., Gabrielli, B., Chan, D., Lees-Miller, S. P., and Lavin, M. F. (1998). ATM associates with and phosphorylates p53: Mapping the region of interaction. *Nature Genet.* **20**, 398–400.

Kirch, S. A., Rathbun, G. A., and Oettinger, M. A. (1998). Dual role of RAG2 in V(D)J recombination: Catalysis and regulation of ordered Ig gene assembly. *EMBO J.* **17**, 4881–4886.

Kishimoto, T. K., Hollander, N., Roberts, T. M., Anderson, D. C., and Springer, T. A. (1987a). Heterogeneous mutations in the beta subunit common to the LFA-1, Mac-1, and p150,95 glycoproteins cause leukocyte adhesion deficiency. *Cell* **50**, 193–202.

Kishimoto, T. K., O'Conner, K., and Springer, T. A. (1989). Leukocyte adhesion deficiency. Aberrant splicing of a conserved integrin sequence causes a moderate deficiency phenotype. *J. Biol. Chem.* **264**, 3588–3595.

Kondo, M., Takeshita, T., Ishii, N., Nakamura, M., Watanabe, S., Arai, K., and Sugamura, K. (1993). Sharing of the interleukin-2 (IL-2) receptor gamma chain between receptors for IL-2 and IL-4. *Science* **262**, 1874–1877.

Korthäuer, U., Graf, D., Mages, H. W., Brière, F., Padayachee, M., Malcolm, S., Ugazio, A. G., Notarangelo, L. D., Levinsky, R. J., and Kroczek, R. A. (1993). Defective expression of T-cell CD40 ligand causes X-linked immunodeficiency with hyper-IgM. *Nature* **361**, 539–541.

Kumar, A., Toscani, A., Rane, S., and Reddy, E. P. (1996). Structural organization and chromosomal mapping of JAK3 locus. *Oncogene* **13**, 2009–2014.

Lamhamedi, S., Jouanguy, E., Altare, F., Roesler, J., and Casanova, J. L. (1998). Interferon-gamma receptor deficiency: Relationship between genotype, environment, and phenotype. *Int. J. Mol. Med.* **1**, 415–418.

Lappalainen, I., Giliani, S., Franceschini, R., Bonnefoy, J. Y., Duckett, C., Notarangelo, L. D., and Vihinen, M. (2000). Structural basis for SH2D1A mutations in X-linked lymphoproliferative disease. *Biochem. Biophys. Res. Commun.* **269**, 124–130.

Leto, T. L., Adams, A. G., and de Mendez, I. (1994). Assembly of the phagocyte NADPH oxidase: Binding of Src homology 3 domains to proline-rich targets. *Proc. Natl. Acad. Sci. (USA)* **91**, 10650–10654.

Leusen, J. H., Bolscher, B. G., Hilarius, P. M., Weening, R. S., Kaulfersch, W., Seger, R. A., Roos, D., and Verhoeven, A. J. (1994b). 156Pro → Gln substitution in the light chain of cytochrome b558 of the human NADPH oxidase (p22-phox) leads to defective translocation of the cytosolic proteins p47-phox and p67-phox. *J. Exp. Med.* **180**, 2329–2334.

Leusen, J. H., de Boer, M., Bolscher, B. G., Hilarius, P. M., Weening, R. S., Ochs, H. D., Roos, D., and Verhoeven, A. J. (1994a). A point mutation in gp91-phox of cytochrome b558 of the human NADPH oxidase leading to defective translocation of the cytosolic proteins p47-phox and p67-phox. *J. Clin. Invest.* **93**, 2120–2126.

Leusen, J. H., de Klein, A., Hilarius, P. M., Åhlin, A., Palmblad, J., Smith, C. I. E., Diekmann, D., Hall, A., Verhoeven, A. J., and Roos, D. (1996). Disturbed interaction of p21-rac with mutated p67-phox causes chronic granulomatous disease. *J. Exp. Med.* **184**, 1243–1249.

Leusen, J. H. W., Meischl, C., Eppink, M. H. M., Hilarius, P. M., de Boer, M., Weening, R. S., Åhlin, A., Sanders, L., Goldblatt, D., Skopczynska, H., Bernatowska, E., Palmblad, J., Verhoeven, A. J., van Berkel, W. J. H., and Roos, D. (2000). Four novel mutations in the gene encoding gp91-*phox* of human NADPH oxidase: Consequences for oxidase assembly. *Blood* **95**, 666–673.

Levin, M., Newport, M. J., D'Souza, S., Kalabalikis, P., Brown, I. N., Lenicker, H. M., Agius, P. V., Davies, E. G., Thrasher, A., Klein, N., and Blackwell, J. M. (1995). Familial disseminated atypical mycobacterial infection in childhood: A human mycobacterial susceptibility gene? *Lancet* **345**, 79–83.

Levy, J., Espanol-Boren, T., Thomas, C., Fischer, A., Tovo, P., Bordigoni, P., Resnick, I., Fasth, A., Baer, M., Gomez, L., Sanders, E. A., Tabone, M. D., Plantaz, D., Etzioni, A., Monafo, V., Abinun, M., Hammarström, L., Abrabamsen, T., Jones, A., Finn, A., Klemola, T., DeVries, E., Sanal, O., Peitsch, M. C., and Notarangelo, L. D. (1997). Clinical spectrum of X-linked hyper-IgM syndrome. *J. Pediatr.* **131**, 47–54.

Li, S. C., Gish, G., Yang, D., Coffey, A. J., Forman-Kay, J. D., Ernberg, I., Kay, L. E., and Pawson, T. (1999). Novel mode of ligand binding by the SH2 domain of the human XLP disease gene product SAP/SH2D1A. *Curr. Biol.* **9**, 1355–1362.

Liu, K. D., Gaffen, S. L., and Goldsmith, M. A. (1998). JAK/STAT signaling by cytokine receptors. *Curr. Opin. Immunol.* **10**, 271–278.

Macchi, P., Villa, A., Gillani, S., Sacco, M. G., Frattini, A., Porta, F., Ugazio, A. G., Johnston, J. A., Candotti, F., O'Shea, J. J., Vezzoni, P., and Notarangelo, L. D. (1995). Mutations of Jak-3 gene in patients with autosomal severe combined immune deficiency (SCID). *Nature* **377**, 65–68.

Machesky, L. M., and Insall, R. H. (1998). Scar1 and the related Wiskott-Aldrich syndrome protein, WASP, regulate the actin cytoskeleton through the Arp2/3 complex. *Curr. Biol.* **8**, 1347–1356.

Masternak, K., Barras, E., Zufferey, M., Conrad, B., Corthals, G., Aebersold, R., Sanchez, J. C., Hochstrasser, D. F., Mach, B., and Reith, W. (1998). A gene encoding a novel RFX-associated transactivator is mutated in the majority of MHC class II deficiency patients. *Nature Genet.* **20**, 273–277.

Matsuda, M., Mayer, B. J., Fukui, Y., and Hanafusa, H. (1990). Binding of transforming protein, P47gag-crk, to a broad range of phosphotyrosine-containing proteins. *Science* **248**, 1537–1539.

Migone, T. S., Lin, J. X., Cereseto, A., Mulloy, J. C., O'Shea, J. J., Franchini, G., and Leonard, W. J. (1995). Constitutively activated Jak-STAT pathway in T cells transformed with HTLV-I. *Science* **269**, 79–81.

Mombaerts, P., Iacomini, J., Johnson, R. S., Herrup, K., Tonegawa, S., and Papaioannou, V. E. (1992). RAG-1-deficient mice have no mature B and T lymphocytes. *Cell* **68**, 869–877.

Moran, M. F., Koch, C. A., Anderson, D., Ellis, C., England, L., Martin, G. S., and Pawson, T. (1990). Src homology region 2 domains direct protein-protein interactions in signal transduction. *Proc. Natl. Acad. Sci. (USA)* **87**, 8622–8626.

Muhlethaler-Mottet, A., Di Berardino, W., Otten, L. A., and Mach, B. (1998). Activation of the MHC class II transactivator CIITA by interferon-gamma requires cooperative interaction between Stat1 and USF-1. *Immunity* **8**, 157–166.

Muhlethaler-Mottet, A., Otten, L. A., Steimle, V., and Mach, B. (1997). Expression of MHC class II molecules in different cellular and functional compartments is controlled by differential usage of multiple promoters of the transactivator CIITA. *EMBO J.* **16**, 2851–2860.

Nagarajan, U. M., Louis-Plence, P., DeSandro, A., Nilsen, R., Bushey, A., and Boss, J. M. (1999). RFX-B is the gene responsible for the most common cause of the bare lymphocyte syndrome, an MHC class II immunodeficiency. *Immunity* **10**, 153–162.

Newport, M. J., Huxley, C. M., Huston, S., Hawrylowicz, C. M., Oostra, B. A., Williamson, R., and Levin, M. (1996). A mutation in the interferon-gamma-receptor gene and susceptibility to mycobacterial infection. *N. Engl. J. Med.* **335**, 1941–1949.

Nichols, K. E., Harkin, D. P., Levitz, S., Krainer, M., Kolquist, K. A., Genovese, C., Bernard, A., Ferguson, M., Zuo, L., Snyder, E., Buckler, A. J., Wise, C., Ashley, J., Lovett, M., Valentine, M. B., Look, A. T., Gerald, W., Housman, D. E., and Haber, D. A. (1998). Inactivating mutations in an SH2 domain-encoding gene in X-linked lymphoproliferative syndrome. *Proc. Natl. Acad. Sci. (USA)* **95**, 13765–13770.

Nisimoto, Y., Motalebi, S., Han, C. H., and Lambeth, J. D. (1999). The p67(phox) activation domain regulates electron flow from NADPH to flavin in flavocytochrome b(558). *J. Biol. Chem.* **274,** 22999–23005.

Noack, D., Heyworth, P. G., Curnutte, J. T., Rae, J., and Cross, A. R. (1999). A novel mutation in the CYBB gene resulting in an unexpected pattern of exon skipping and chronic granulomatous disease. *Biochim. Biophys. Acta* **1454,** 270–274.

Noguchi, M., Yi, H., Rosenblatt, H. M., Filipovich, A. H., Adelstein, S., Modi, W. S., McBride, O. W., and Leonard, W. J. (1993a). Interleukin-2 receptor gamma chain mutation results in X-linked severe combined immunodeficiency in humans. *Cell* **73,** 147–157.

Notarangelo, L. D., Peitsch, M. C., Abrahamsen, T. G., Bachelot, C., Bordigoni, P., Cant, A. J., Chapel, H., Clementi, M., Deacock, S., de Saint Basile, G., Duse, M., Espanol, T., Etzioni, A., Fasth, A., Fischer, A., Giliani, S., Gomez, L., Hammarström, L., Jones, A., Kanariou, M., Kinnon, C., Klemola, T., Kroczek, R. A., Levy, J., Matamoros, N., Monafo, V., Paolucci, P., Reznick, I., Sanal, O., Smith, C. I. E., Thompson, R. A., Tovo, P., Villa, A., Vihinen, M., Vossen, J., and Zegers, B. J. M. (1996). CD40Lbase: A database of CD40L gene mutations causing X-linked hyper-IgM syndrome. *Immunol. Today* **17,** 511–516.

Ochs, H. D. (1998). The Wiskott-Aldrich syndrome. *Springer Semin. Immunopathol.* **19,** 435–458.

Ochs, H. D., Smith, C. I. E., and Puck, J. M. (eds.) (1999). Primary Immunodeficiency Diseases. A Molecular and Genetic Approach pp. 1–501. Oxford University Press, New York, Oxford.

Oettinger, M. A., Schatz, D. G., Gorka, C., and Baltimore, D. (1990). RAG-1 and RAG-2, adjacent genes that synergistically activate V(D)J recombination. *Science* **248,** 1517–1523.

Onodera, M., Nelson, D. M., Sakiyama, Y., Candotti, F., and Blaese, R. M. (1999). Gene therapy for severe combined immunodeficiency caused by adenosine deaminase deficiency: Improved retroviral vectors for clinical trials. *Acta Haematol.* **101,** 89–96.

O'Shea, J. J. (1997). Jaks, STATs, cytokine signal transduction, and immunoregulation: Are we there yet? *Immunity* **7,** 1–11.

Pan-Yun Ting, J., and Zhu, X. S. (1999). Class II MHC genes: A model gene regulatory system with great biologic consequences. *Microbes Infect.* **1,** 855–861.

Parolini, O., Berardelli, S., Riedl, E., Bello-Fernandez, C., Strobl, H., Majdic, O., and Knapp, W. (1997). Expression of Wiskott-Aldrich syndrome protein (WASP) gene during hematopoietic differentiation. *Blood* **90,** 70–75.

Parolini, O., Ressmann, G., Haas, O. A., Pawlowsky, J., Gadner, H., Knapp, W., and Holter, W. (1998). X-linked Wiskott-Aldrich syndrome in a girl. *N. Engl. J. Med.* **338,** 291–295.

Peijnenburg, A., Van den Berg, R., Van Eggermond, M. J. C. A., Sanal, O., Vossen, J. M. J. J., Lennon, A.-M. C., A.-L., and Van den Elsen, P. (2000). Defective MHC class II expression in an MHC class II deficient patient is caused by a novel deletion of a splice donor site in the MHC class II transactivator gene. *Immunogenetics,* **51,** 42–49.

Peijnenburg, A., Van Eggermond, M. C., Van den Berg, R., Sanal, O., Vossen, J. M., and Van den Elsen, P. J. (1999a). Molecular analysis of an MHC class II deficiency patient reveals a novel mutation in the RFX5 gene. *Immunogenetics* **49,** 338–345.

Peijnenburg, A., Van Eggermond, M. J., Gobin, S. J., Van den Berg, R., Godthelp, B. C., Vossen, J. M., and Van den Elsen, P. J. (1999b). Discoordinate expression of invariant chain and MHC class II genes in class II transactivator-transfected fibroblasts defective for RFX5. *J. Immunol.* **163,** 794–801.

Petrella, A., Doti, I., Agosti, V., Giarrusso, P. C., Vitale, D., Bond, H. M., Cuomo, C., Tassone, P., Franco, B., Ballabio, A., Venuta, S., and Morrone, G. (1998). A 5′ regulatory sequence containing two Ets motifs controls the expression of the Wiskott-Aldrich syndrome protein (WASP) gene in human hematopoietic cells. *Blood* **91,** 4554–4560.

Pierre-Audigier, C., Jouanguy, E., Lamhamedi, S., Altare, F., Rauzier, J., Vincent, V., Canioni, D., Emile, J. F., Fischer, A., Blanche, S., Gaillard, J. L., and Casanova, J. L. (1997). Fatal disseminated

Mycobacterium smegmatis infection in a child with inherited interferon gamma receptor deficiency. *Clin. Infect. Dis.* **24,** 982–984.

Porter, C. D., Parkar, M. H., Verhoeven, A. J., Levinsky, R. J., Collins, M. K., and Kinnon, C. (1994). p22-phox-Deficient chronic granulomatous disease: Reconstitution by retrovirus-mediated expression and identification of a biosynthetic intermediate of gp91-phox. *Blood* **84,** 2767–2775.

Poy, F., Yaffe, M. B., Sayos, J., Saxena, K., Morra, M., Sumegi, J., Cantley, L. C., Terhorst, C., and Eck, M. J. (1999). Crystal structures of the XLP protein SAP reveal a class of SH2 domains with extended, phosphotyrosine-independent sequence recognition. *Mol. Cell* **4,** 555–561.

Prehoda, K. E., Lee, D. J., and Lim, W. A. (1999). Structure of the enabled/VASP homology 1 domain-peptide complex: A key component in the spatial control of actin assembly. *Cell* **97,** 471–480.

Puck, J. M. (1999). X-linked severe combined immunodeficiency. *In* "Primary Immunodeficiency Diseases. A Molecular and Genetic Approach" (H. D. Ochs, C. I. E. Smith, and J. M. Puck, eds.), pp. 99–110. Oxford University Press, New York, Oxford.

Puck, J. M., de Saint Basile, G., Schwarz, K., Fugmann, S., and Fischer, R. E. (1996). IL2RGbase: A database of gamma c-chain defects causing human X-SCID. *Immunol. Today* **17,** 507–511.

Puck, J. M., Pepper, A. E., Henthorn, P. S., Candotti, F., Isakov, J., Whitwam, T., Conley, M. E., Fischer, R. E., Rosenblatt, H. M., Small, T. N., and Buckley, R. H. (1997). Mutation analysis of IL2RG in human X-linked severe combined immunodeficiency. *Blood* **89,** 1968–1977.

Purtilo, D. T., Cassel, C., and Yang, J. P. (1974). Fatal infectious mononucleosis in familial lympho-histiocytosis. *N. Engl. J. Med.* **291,** 736.

Quan, V., Towey, M., Sacks, S., and Kelly, A. P. (1999). Absence of MHC class II gene expression in a patient with a single amino acid substitution in the class II transactivator protein CIITA. *Immunogenetics* **49,** 957–963.

Quie, P. G., White, J. G., Holmes, B., and Good, R. A. (1967). In vitro bactericidal capacity of human polymorphonuclear leukocytes: Diminished activity in chronic granulomatous disease of childhood. *J. Clin. Invest.* **46,** 668–679.

Rane, S. G., and Reddy, E. P. (1994). JAK3: A novel JAK kinase associated with terminal differentiation of hematopoietic cells. *Oncogene* **9,** 2415–2423.

Rawlings, D. J., Saffran, D. C., Tsukada, S., Largaespada, D. A., Grimaldi, J. C., Cohen, L., Mohr, R. N., Bazan, J. F., Howard, M., Copeland, N. G., Jenkins, N. A., and Witte, O. N. (1993). Mutation of unique region of Bruton's tyrosine kinase in immunodeficient XID mice. *Science* **261,** 358–361.

Rawlings, S. L., Crooks, G. M., Bockstoce, D., Barsky, L. W., Parkman, R., and Weinberg, K. I. (1999). Spontaneous apoptosis in lymphocytes from patients with Wiskott-Aldrich syndrome: Correlation of accelerated cell death and attenuated bcl-2 expression. *Blood* **94,** 3872–3882.

Reith, W., Muhlethaler-Mottet, A., Masternak, K., Villard, J., and Mach, B. (1999a). The molecular basis of MHC class II deficiency and transcriptional control of MHC class II gene expression. *Microbes Infect.* **1,** 839–846.

Reith, W., Steimle, V., Lisowska-Grospierre, B., A., F., and Mach, B. (1999b). Molecular basis of major histocompatibility complex class II deficiency. *In* "Primary Immunodeficiency Diseases. A Molecular and Genetic Approach" (H. D. Ochs, C. I. E. Smith, and J. M. Puck, eds.), pp. 167–180. Oxford University Press, New York, Oxford.

Remold-O'Donnell, E., Rosen, F. S., and Kenney, D. M. (1996). Defects in Wiskott-Aldrich syndrome blood cells. *Blood* **87,** 2621–2631.

Riedy, M. C., Dutra, A. S., Blake, T. B., Modi, W., Lal, B. K., Davis, J., Bosse, A., O'Shea, J. J., and Johnston, J. A. (1996). Genomic sequence, organization, and chromosomal localization of human JAK3. *Genomics* **37,** 57–61.

Riikonen, P., and Vihinen, M. (1999). MUTbase: Maintenance and analysis of distributed mutation databases. *Bioinformatics* **15,** 852–859.

Rivera-Matos, I. R., Rakita, R. M., Mariscalco, M. M., Elder, F. F., Dreyer, S. A., and Cleary, T. G. (1995). Leukocyte adhesion deficiency mimicking Hirschsprung disease. *J. Pediatr.* **127,** 755–757.

Roesler, J., Kofink, B., Wendisch, J., Heyden, S., Paul, D., Friedrich, W., Casanova, J. L., Leupold, W., Gahr, M., and Rösen-Wolff, A. (1999). Listeria monocytogenes and recurrent mycobacterial infections in a child with complete interferon-gamma-receptor (IFNgammaR1) deficiency: Mutational analysis and evaluation of therapeutic options. *Exp. Hematol.* **27,** 1368–1374.

Roos, D., and Curnutte, J. C. (1999). Chronic granulomatous disease. In "Primary Immunodeficiency Diseases. A Molecular and Genetic Approach" (H. D. Ochs, C. I. E. Smith, and J. M. Puck, eds.), pp. 353–374. Oxford University Press, New York, Oxford.

Roos, D., Curnutte, J. C., Hossle, J. P., Lau, Y. L., Ariga, T., Nunoi, H., Dinauer, M. C., Gahr, M., Segal, A. W., Newburger, P. E., Giacca, M., Keep, N. H., and van Zwieten, R. (1996a). X-CGDbase: A database of X-CGD-causing mutations. *Immunol. Today* **17,** 517–521.

Roos, D., de Boer, M., Kuribayashi, F., Meischl, C., Weening, R. S., Segal, A. W., Ahlin, A., Nemet, K., Hossle, J. P., Bernatowska-Matuszkiewicz, E., and Middleton-Price, H. (1996b). Mutations in the X-linked and autosomal recessive forms of chronic granulomatous disease. *Blood* **87,** 1663–1681.

Royer-Pokora, B., Kunkel, L. M., Monaco, A. P., Goff, S. C., Newburger, P. E., Baehner, R. L., Cole, F. S., Curnutte, J. T., and Orkin, S. H. (1986). Cloning the gene for an inherited human disorder–chronic granulomatous disease—on the basis of its chromosomal location. *Nature* **322,** 32–38.

Russell, S. M., Johnston, J. A., Noguchi, M., Kawamura, M., Bacon, C. M., Friedmann, M., Berg, M., McVicar, D. W., Witthuhn, B. A., Silvennoinen, O., Goldman, A. S., Schmalstieg, F. C., Ihle, J. N., O'Shea, J. J., and Leonard, W. J. (1994). Interaction of IL-2R beta and gamma c chains with Jak1 and Jak3: Implications for XSCID and XCID. *Science* **266,** 1042–1045.

Russell, S. M., Keegan, A. D., Harada, N., Nakamura, Y., Noguchi, M., Leland, P., Friedmann, M. C., Miyajima, A., Puri, R. K., Paul, W. E., and Leonard, W. J. (1993). Interleukin-2 receptor gamma chain: A functional component of the interleukin-4 receptor. *Science* **262,** 1880–1883.

Russell, S. M., Tayebi, N., Nakajima, H., Riedy, M. C., Roberts, J. L., Aman, M. J., Migone, T. S., Noguchi, M., Markert, M. L., Buckley, R. H., O'Shea, J. J., and Leonard, W. J. (1995). Mutation of Jak3 in a patient with SCID: essential role of Jak3 in lymphoid development. *Science* **270,** 797–800.

Safford, M. G., Levenstein, M., Tsifrina, E., Amin, S., Hawkins, A. L., Griffin, C. A., Civin, C. I., and Small, D. (1997). JAK3: Expression and mapping to chromosome 19p12-13.1. *Exp. Hematol.* **25,** 374–386.

Santagata, S., Besmer, E., Villa, A., Bozzi, F., Allingham, J. S., Sobacchi, C., Haniford, D. B., Vezzoni, P., Nussenzweig, M. C., Pan, Z. Q., and Cortes, P. (1999). The RAG1/RAG2 complex constitutes a 3′ flap endonuclease: Implications for junctional diversity in V(D)J and transpositional recombination. *Mol. Cell* **4,** 935–947.

Santisteban, I., Arredondo-Vega, F. X., Kelly, S., Mary, A., Fischer, A., Hummell, D. S., Lawton, A., Sorensen, R. U., Stiehm, E. R., Uribe, L., Weinberg, K., and Hershfield, M. S. (1993). Novel splicing, missense, and deletion mutations in seven adenosine deaminase-deficient patients with late/delayed onset of combined immunodeficiency disease. Contribution of genotype to phenotype. *J. Clin. Invest.* **92,** 2291–2302.

Savitsky, K., Bar-Shira, A., Gilad, S., Rotman, G., Ziv, Y., Vanagaite, L., Tagle, D. A., Smith, S., Uziel, T., Sfez, S., Ashkenazi M., Pecker, I., Fryman, M., Harnik, R., Patanjali, S. R., Simmons, A., Clines, G. A., Sartiel, A., Gatti, R. A., Chessa, L., Sanal, O., Lavin, M. F., Jaspers, NGJ, Taylor, AMR, Arlett, C. F., Miki, T., Weissman, S. M., Lovett, M., Collins, F. S., and Shiloh, Y. (1995a). A single ataxia telangiectasia gene with a product similar to PI-3 kinase. *Science* **268,** 1749–1753.

Savitsky, K., Sfez, S., Tagle, D. A., Ziv, Y., Sartiel, A., Collins, F. S., Shiloh, Y., and Rotman, G. (1995b). The complete sequence of the coding region of the ATM gene reveals similarity to cell cycle regulators in different species. *Hum. Mol. Genet.* **4,** 2025–2032.

Sayos, J., Wu, C., Morra, M., Wang, N., Zhang, X., Allen, D., van Schaik, S., Notarangelo, L., Geha, R., Roncarolo, M. G., Oettgen, H., De Vries, J. E., Aversa, G., and Terhorst, C. (1998). The X-linked lymphoproliferative-disease gene product SAP regulates signals induced through the co-receptor SLAM. *Nature* **395,** 462–469.

Schatz, D. G., Oettinger, M. A., and Baltimore, D. (1989). The V(D)J recombination activating gene, RAG-1. *Cell* **59,** 1035–1048.

Schumacher, R. F., Mella, P., Badolato, R., Fiorini, M., Savoldi, G., Giliani, S., Villa, A., Candotti, F., Tampalini, A., O'Shea, J. J., and Notarangelo, L. D. (2000). Complete genomic organization of the human JAK3 gene and mutation analysis in severe combined immunodeficiency by single-strand conformation polymorphism. *Hum. Genet.* **106,** 73–79.

Schumacher, R. F., Mella, P., Lalatta, F., Fiorini, M., Giliani, S., Villa, A., Candotti, F., and Notarangelo, L. D. (1999). Prenatal diagnosis of JAK3 deficient SCID. *Prenat. Diagn.* **19,** 653–656.

Schuster, V., and Kreth, H. W. (1999). X-linked lymphoproliferative disease. *In* "Primary Immunodeficiency Diseases. A Molecular and Genetic Approach" (H. D. Ochs, C. I. E. Smith, and J. M. Puck, eds.), pp. 222–232. Oxford University Press, New York, Oxford.

Schwarz, K., Gauss, G. H., Ludwig, L., Pannicke, U., Li, Z., Lindner, D., Friedrich, W., Seger, R. A., Hansen-Hagge, T. E., Desiderio, S., Lieber, M. R., and Bartram, C. R. (1996a). RAG mutations in human B cell-negative SCID. *Science* **274,** 97–99.

Schwarz, K., Nonoyama, S., Peitsch, M. C., de Saint Basile, G., Espanol, T., Fasth, A., Fischer, A., Freitag, K., Friedrich, W., Fugmann, S., Hossle, H. P., Jones, A., Kinnon, C., Meindl, A., Notarangelo, L. D., Wechsler, A., Weiss, M., and Ochs, H. D. (1996b). WASPbase: A database of WAS- and XLT-causing mutations. *Immunol. Today* **17,** 496–502.

Seemayer, T. A., Gross, T. G., Egeler, R. M., Pirruccello, S. J., Davis, J. R., Kelly, C. M., Okano, M., Lanyi, A., and Sumegi, J. (1995). X-linked lymphoproliferative disease: Twenty-five years after the discovery. *Pediatr. Res.* **38,** 471–478.

Segal, A. W., Heyworth, P. G., Cockcroft, S., and Barrowman, M. M. (1985). Stimulated neutrophils from patients with autosomal recessive chronic granulomatous disease fail to phosphorylate a M_r-44,000 protein. *Nature* **316,** 547–549.

Segal, A. W., West, I., Wientjes, F., Nugent, J. H., Chavan, A. J., Haley, B., Garcia, R. C., Rosen, H., and Scrace, G. (1992). Cytochrome b-245 is a flavocytochrome containing FAD and the NADPH-binding site of the microbicidal oxidase of phagocytes. *Biochem. J.* **284,** 781–788.

Seyama, K., Nonoyama, S., Gangsaas, I., Hollenbaugh, D., Pabst, H. F., Aruffo, A., and Ochs, H. D. (1998). Mutations of the CD40 ligand gene and its effect on CD40 ligand expression in patients with X-linked hyper IgM syndrome. *Blood* **92,** 2421–2434.

Shafman, T., Khanna, K. K., Kedar, P., Spring, K., Kozlov, S., Yen, T., Hobson, K., Gatei, M., Zhang, N., Watters, D., Egerton, M., Shiloh, Y., Kharbanda, S., Kufe, D., and Lavin, M. F. (1997). Interaction between ATM protein and c-Abl in response to DNA damage. *Nature* **387,** 520–523.

Shimadzu, M., Nunoi, H., Terasaki, H., Ninomiya, R., Iwata, M., Kanegasaka, S., and Matsuda, I. (1995). Structural organization of the gene for CD40 ligand: Molecular analysis for diagnosis of X-linked hyper-IgM syndrome. *Biochim. Biophys. Acta* **1260,** 67–72.

Shinkai, Y., Rathbun, G., Lam, K. P., Oltz, E. M., Stewart, V., Mendelsohn, M., Charron, J., Datta, M., Young, F., Stall, A. M., and Alt, F. W. (1992). RAG-2-deficient mice lack mature lymphocytes owing to inability to initiate V(D)J rearrangement. *Cell* **68,** 855–867.

Sideras, P., and Smith, C. I. E. (1995). Molecular and cellular aspects of X-linked agammaglobulinemia. *Adv. Immunol.* **59,** 135–223.

Signorini, S., Imberti, L., Pirovano, S., Villa, A., Facchetti, F., Ungari, M., Bozzi, F., Albertini, A., Ugazio, A. G., Vezzoni, P., and Notarangelo, L. D. (1999). Intrathymic restriction and peripheral expansion of the T-cell repertoire in Omenn syndrome. *Blood* **94,** 3468–3478.

Smith, C. I. E., Baskin, B., Humire-Greiff, P., Zhou, J. N., Olsson, P. G., Maniar, H. S., Kjellén, P., Lambris, J. D., Christensson, B., Hammarström, L., Bentley, D., Vetrie, D., Islam, K. B., Vorechovsky, I., and Sideras, P. (1994). Expression of Bruton's agammaglobulinemia tyrosine kinase gene, BTK, is selectively down-regulated in T lymphocytes and plasma cells. *J. Immunol.* **152,** 557–565.

Smith, C. I. E., and Notarangelo, L. D. (1997). Molecular basis for X-linked immunodeficiencies. *Adv. Genet.* **35,** 57–115.

Smith, C. I. E., and Vihinen, M. (1996). Immunodeficiency mutation databases—A new research tool. *Immunol. Today* **17**, 495–496.

Snapper, S. B., Rosen, F. S., Mizoguchi, E., Cohen, P., Khan, W., Liu, C. H., Hagemann, T. L., Kwan, S. P., Ferrini, R., Davidson, L., Bhan, A. K., and Alt, F. W. (1998). Wiskott-Aldrich syndrome protein-deficient mice reveal a role for WASP in T but not B cell activation. *Immunity* **9**, 81–91.

Solomon, E., Palmer, R. W., Hing, S., and Law, S. K. (1988). Regional localization of CD18, the beta-subunit of the cell surface adhesion molecule LFA-1, on human chromosome 21 by in situ hybridization. *Ann. Hum. Genet.* **52**, 123–128.

Steimle, V., Durand, B., Barras, E., Zufferey, M., Hadam, M. R., Mach, B., and Reith, W. (1995). A novel DNA-binding regulatory factor is mutated in primary MHC class II deficiency (bare lymphocyte syndrome). *Genes Dev.* **9**, 1021–1032.

Steimle, V., Otten, L. A., Zufferey, M., and Mach, B. (1993). Complementation cloning of an MHC class II transactivator mutated in hereditary MHC class II deficiency (or bare lymphocyte syndrome). *Cell* **75**, 135–146.

Stephan, V., Wahn, V., Le Deist, F., Dirksen, U., Broker, B., Müller-Fleckenstein, I., Horneff, G., Schroten, H., Fischer, A., and de Saint Basile, G. (1996). Atypical X-linked severe combined immunodeficiency due to possible spontaneous reversion of the genetic defect in T cells. *N. Engl. J. Med.* **335**, 1563–1567.

Stewart, D. M., Treiber-Held, S., Kurman, C. C., Facchetti, F., Notarangelo, L. D., and Nelson, D. L. (1996). Studies of the expression of the Wiskott-Aldrich syndrome protein. *J. Clin. Invest.* **97**, 2627–2634.

Sullivan, J. L., and Woda, B. A. (1989). X-linked lymphoproliferative syndrome. *Immunodefic. Rev.* **1**, 325–347.

Sumimoto, H., Kage, Y., Nunoi, H., Sasaki, H., Nose, T., Fukumaki, Y., Ohno, M., Minakami, S., and Takeshige, K. (1994). Role of Src homology 3 domains in assembly and activation of the phagocyte NADPH oxidase. *Proc. Natl. Acad. Sci. (USA)* **91**, 5345–5349.

Suzuki, S., Kumatori, A., Haagen, I. A., Fujii, Y., Sadat, M. A., Jun, H. L., Tsuji, Y., Roos, D., and Nakamura, M. (1998). PU.1 as an essential activator for the expression of gp91(phox) gene in human peripheral neutrophils, monocytes, and B lymphocytes. *Proc. Natl. Acad. Sci. (USA)* **95**, 6085–6090.

Takahashi, T., and Shirasawa, T. (1994). Molecular cloning of rat JAK3, a novel member of the JAK family of protein tyrosine kinases. *FEBS Lett.* **342**, 124–128.

Takemoto, S., Mulloy, J. C., Cereseto, A., Migone, T. S., Patel, B. K., Matsuoka, M., Yamaguchi, K., Takatsuki, K., Kamihira, S., White, J. D., Leonard, W. J., Waldmann, T., and Franchini, G. (1997). Proliferation of adult T cell leukemia/lymphoma cells is associated with the constitutive activation of JAK/STAT proteins. *Proc. Natl. Acad. Sci. (USA)* **94**, 13897–13902.

Tangye, S. G., Lazetic, S., Woollatt, E., Sutherland, G. R., Lanier, L. L., and Phillips, J. H. (1999). Cutting edge: Human 2B4, an activating NK cell receptor, recruits the protein tyrosine phosphatase SHP-2 and the adaptor signaling protein SAP. *J. Immunol.* **162**, 6981–6985.

Taylor, W. R., Jones, D. T., and Segal, A. W. (1993). A structural model for the nucleotide binding domains of the flavocytochrome b-245 beta-chain. *Protein Sci.* **2**, 1675–1685.

Teahan, C., Rowe, P., Parker, P., Totty, N., and Segal, A. W. (1987). The X-linked chronic granulomatous disease gene codes for the beta-chain of cytochrome b-245. *Nature* **327**, 720–721.

Telatar, M., Teraoka, S., Wang, Z., Chun, H. H., Liang, T., Castellvi-Bel, S., Udar, N., Borresen-Dale, A. L., Chessa, L., Bernatowska-Matuszkiewicz, E., Porras, O., Watanabe, M., Junker, A., Concannon, P., and Gatti, R. A. (1998). Ataxia-telangiectasia: Identification and detection of founder-effect mutations in the ATM gene in ethnic populations. *Am. J. Hum. Genet.* **62**, 86–97.

Thomas, J. D., Sideras, P., Smith, C. I., Vořechovský, I., Chapman, V., and Paul, W. E. (1993). Colocalization of X-linked agammaglobulinemia and X-linked immunodeficiency genes. *Science* **261**, 355–358.

Thompson, A. D., Braun, B. S., Arvand, A., Stewart, S. D., May, W. A., Chen, E., Korenberg, J., and Denny, C. (1996). EAT-2 is a novel SH2 domain containing protein that is up regulated by Ewing's sarcoma EWS/FLI1 fusion gene. *Oncogene* **13**, 2649–2658.

Tortolani, P. J., Lal, B. K., Riva, A., Johnston, J. A., Chen, Y. Q., Reaman, G. H., Beckwith, M., Longo, D., Ortaldo, J. R., Bhatia, K., McGrath, I., Kehrl, J., Tuscano, J., McVicar, D. W., and O'Shea, J. J. (1995). Regulation of JAK3 expression and activation in human B cells and B cell malignancies. *J. Immunol.* **155**, 5220–5226.

Tsukada, S., Saffran, D. C., Rawlings, D. J., Parolini, O., Allen, R. C., Klisak, I., Sparkes, R. S., Kubagawa, H., Mohandas, T., Quan, S., Belmont, J. W., Cooper, M. D., Conley, M. E., and Witte, O. N. (1993). Deficient expression of a B cell cytoplasmic tyrosine kinase in human X-linked agammaglobulinemia. *Cell* **72**, 279–290.

Verbsky, J. W., Bach, E. A., Fang, Y. F., Yang, L., Randolph, D. A., and Fields, L. E. (1996). Expression of Janus kinase 3 in human endothelial and other non-lymphoid and non-myeloid cells. *J. Biol. Chem.* **271**, 13976–13980.

Vetrie, D., Vořechovský, I., Sideras, P., Holland, J., Davies, A., Flinter, F., Hammarström, L., Kinnon, C., Levinsky, R., Bobrow, M., Smith, C. I. E., and Bentley, D. R. (1993). The gene involved in X-linked agammaglobulinaemia is a member of the src family of protein-tyrosine kinases. *Nature* **361**, 226–233.

Vihinen, M., Brooimans, R. A., Kwan, S. P., Leväslaiho, H., Litman, G. W., Ochs, H. D., Resnick, I., Schwaber, J. H., Vořechovský, I., and Smith, C. I. E. (1996b). BTKbase: XLA – mutation registry. *Immunol. Today* **17**, 502–506.

Vihinen, M., Cooper, M. D., de Saint Basile, G., Fischer, A., Good, R. A., Hendriks, R. W., Kinnon, C., Kwan, S. P., Litman, G. W., Notarangelo, L. D., Ochs, H. D., Rosen, F. S., Vertrie, D., Webster, A. D. B., Zegers, B. J. M., and Smith, C. I. E. (1995). BTKbase: A database of XLA-causing mutations. *Immunol. Today* **16**, 460–465.

Vihinen, M., Iwata, T., Kinnon, C., Kwan, S. P., Ochs, H. D., Vořechovský, I., and Smith, C. I. E. (1996a). BTKbase, mutation database for X-linked agammaglobulinemia (XLA). *Nucleic Acids Res.* **24**, 160–165.

Vihinen, M., Kwan, S. P., Lester, T., Ochs, H. D., Resnick, I., Väliaho, J., Conley, M. E., and Smith, C. I. E. (1999b). Mutations of the human BTK gene coding for bruton tyrosine kinase in X-linked agammaglobulinemia. *Hum. Mutat.* **13**, 280–285.

Vihinen, M., Leväslaiho, H., and Cotton, R. D. (1999a). Immunodeficiency mutation databases. *In* "Primary Immunodeficiency Diseases. A Molecular and Genetic Approach" (H. D. Ochs, C. I. E. Smith, and M. Puck, eds.), pp. 443–447. Oxford University Press, New York, Oxford.

Vihinen, M., Mattsson, P. T., and Smith, C. I. E. (1997a). BTK, the tyrosine kinase affected in X-linked agammaglobulinemia. *Front. Biosci.* **2**, d27–d42.

Vihinen, M., Nilsson, L., and Smith, C. I. E. (1994b). Structural basis of SH2 domain mutations in X-linked agammaglobulinemia. *Biochem. Biophys. Res. Commun.* **205**, 1270–1277.

Vihinen, M., Nore, B. F., Mattsson, P. T., Bäckesjö, C. M., Nars, M., Koutaniemi, S., Watanabe, C., Lester, T., Jones, A., Ochs, H. D., and Smith, C. I. E. (1997b). Missense mutations affecting a conserved cysteine pair in the TH domain of Btk. *FEBS Lett.* **413**, 205–210.

Vihinen, M., Vetrie, D., Maniar, H. S., Ochs, H. D., Zhu, Q., Vořechovský, I., Webster, A. D., Notarangelo, L. D., Nilsson, L., Sowadski, J. M., and Smith, C. I. E. (1994a). Structural basis for chromosome X-linked agammaglobulinemia: a tyrosine kinase disease. *Proc. Natl. Acad. Sci. (USA)* **91**, 12803–12807.

Villa, A., Notarangelo, L. D., Di Santo, J. P., Macchi, P. P., Strina, D., Frattini, A., Lucchini, F., Patrosso, C. M., Giliani, S., Mantuano, E., Agosti, S., Nocera, G., Kroczek, R. A., Fischer, A., Ugazio, A. G., de Saint Basile, G., and Vezzoni, P. (1994). Organization of the human CD40L gene: Implications for molecular defects in X chromosome-linked hyper-IgM syndrome and prenatal diagnosis. *Proc. Natl. Acad. Sci. (USA)* **91**, 2110–2114.

Villa, A., Santagata, S., Bozzi, F., Giliani, S., Frattini, A., Imberti, L., Gatta, L. B., Ochs, H. D., Schwarz, K., Notarangelo, L. D., Vezzoni, P., and Spanopoulou, E. (1998). Partial V(D)J recombination activity leads to Omenn syndrome. *Cell* **93**, 885–896.

Villa, A., Sironi, M., Macchi, P., Matteucci, C., Notarangelo, L. D., Vezzoni, P., and Mantovani, A. (1996). Monocyte function in a severe combined immunodeficient patient with a donor splice site mutation in the Jak3 gene. *Blood* **88**, 817–823.

Villard, J., Lisowska-Grospierre, B., van den Elsen, P., Fischer, A., Reith, W., and Mach, B. (1997b). Mutation of RFXAP, a regulator of MHC class II genes, in primary MHC class II deficiency. *N. Engl. J. Med.* **337**, 748–753.

Villard, J., Reith, W., Barras, E., Gos, A., Morris, M. A., Antonarakis, S. E., Van den Elsen, P. J., and Mach, B. (1997a). Analysis of mutations and chromosomal localisation of the gene encoding RFX5, a novel transcription factor affected in major histocompatibility complex class II deficiency. *Hum. Mutat.* **10**, 430–435.

Vořechovský, I., Luo, L., Dyer, M. J., Catovsky, D., Amlot, P. L., Yaxley, J. C., Foroni, L., Hammarström, L., Webster, A. D., and Yuille, M. A. (1997). Clustering of missense mutations in the ataxia-telangiectasia gene in a sporadic T-cell leukaemia. *Nature Genet.* **17**, 96–99.

Vořechovský, I., Webster, A. D. B., Lähdesmäki, A., Plebani, A., and Hammarström, L. (1999). Mapping chromosome susceptibility loci in human immunoglobulin A (IgA) deficiency: Why do A-T patients lack IgA? Abstract. A-T meeting, Las Vegas, USA.

Wallach, T. M., and Segal, A. W. (1997). Analysis of glycosylation sites on gp91phox, the flavocytochrome of the NADPH oxidase, by site-directed mutagenesis and translation in vitro. *Biochem. J.* **321**, 583–585.

Wiginton, D. A., Adrian, G. S., and Hutton, J. J. (1984). Sequence of human adenosine deaminase cDNA including the coding region and a small intron. *Nucleic Acids Res.* **12**, 2439–2446.

Wiginton, D. A., Kaplan, D. J., States, J. C., Akeson, A. L., Perme, C. M., Bilyk, I. J., Vaughn, A. J., Lattier, D. L., and Hutton, J. J. (1986). Complete sequence and structure of the gene for human adenosine deaminase. *Biochemistry* **25**, 8234–8244.

Williams, G. S., Malin, M., Vremec, D., Chang, C. H., Boyd, R., Benoist, C., and Mathis, D. (1998). Mice lacking the transcription factor CIITA—A second look. *Int. Immunol.* **10**, 1957–1967.

Wilson, D. K., and Quiocho, F. A. (1993). A pre-transition-state mimic of an enzyme: X-ray structure of adenosine deaminase with bound 1-deazaadenosine and zinc-activated water. *Biochemistry* **32**, 1689–1694.

Wilson, D. K., Rudolph, F. B., and Quiocho, F. A. (1991). Atomic structure of adenosine deaminase complexed with a transition-state analog: Understanding catalysis and immunodeficiency mutations. *Science* **252**, 1278–1284.

Witthuhn, B. A., Silvennoinen, O., Miura, O., Lai, K. S., Cwik, C., Liu, E. T., and Ihle, J. N. (1994). Involvement of the Jak-3 Janus kinase in signalling by interleukins 2 and 4 in lymphoid and myeloid cells. *Nature* **370**, 153–157.

Wright, A. H., Douglass, W. A., Taylor, G. M., Lau, Y. L., Higgins, D., Davies, K. A., and Law, S. K. (1995). Molecular characterization of leukocyte adhesion deficiency in six patients. *Eur. J. Immunol.* **25**, 717–722.

Xu, G. L., Bestor, T. H., Bourc'his, D., Hsieh, C. L., Tommerup, N., Bugge, M., Hulten, M., Qu, X., Russo, J. J., and Viegas-Pequignot, E. (1999). Chromosome instability and immunodeficiency syndrome caused by mutations in a DNA methyltransferase gene. *Nature* **402**, 187–191.

Xu, X., Kang, S. H., Heidenreich, O., Okerholm, M., O'Shea, J. J., and Nerenberg, M. I. (1995). Constitutive activation of different Jak tyrosine kinases in human T cell leukemia virus type 1 (HTLV-1) tax protein or virus-transformed cells. *J. Clin. Invest.* **96**, 1548–1555.

Xu, Y., Ashley, T., Brainerd, E. E., Bronson, R. T., Meyn, M. S., and Baltimore, D. (1996). Targeted disruption of ATM leads to growth retardation, chromosomal fragmentation during meiosis, immune defects, and thymic lymphoma. *Genes Dev.* **10**, 2411–2422.

Yin, L., Ferrand, V., Lavoue, M. F., Hayoz, D., Philippe, N., Souillet, G., Seri, M., Giacchino, R., Castagnola, E., Hodgson, S., Sylla, B. S., and Romeo, G. (1999). SH2D1A mutation analysis for diagnosis of XLP in typical and atypical patients. *Hum. Genet.* **105**, 501–505.

Yoo, J., Stone, R. T., Solinas-Toldo, S., Fries, R., and Beattie, C. W. (1996). Cloning and chromosomal mapping of bovine interleukin-2 receptor gamma gene. *DNA Cell Biol.* **15**, 453–459.

Yoshimura, A. (1998). The CIS/JAB family: Novel negative regulators of JAK signaling pathways. *Leukemia* **12**, 1851–1857.

Yu, L., Zhen, L., and Dinauer, M. C. (1997). Biosynthesis of the phagocyte NADPH oxidase cytochrome b558. Role of heme incorporation and heterodimer formation in maturation and stability of gp91phox and p22phox subunits. *J. Biol. Chem.* **272**, 27288–27294.

Zhang, J., Shehabeldin, A., da Cruz, L. A., Butler, J., Somani, A. K., McGavin, M., Kozieradzki, I., dos Santos, A. O., Nagy, A., Grinstein, S., Penninger, J. M., and Siminovitch, K. A. (1999). Antigen receptor-induced activation and cytoskeletal rearrangement are impaired in Wiskott-Aldrich syndrome protein-deficient lymphocytes. *J. Exp. Med.* **190**, 1329–1342.

Zhou, Y. J., Hanson, E. P., Chen, Y. Q., Magnuson, K., Chen, M., Swann, P. G., Wange, R. L., Changelian, P. S., and O'Shea, J. J. (1997). Distinct tyrosine phosphorylation sites in JAK3 kinase domain positively and negatively regulate its enzymatic activity. *Proc. Natl. Acad. Sci. (USA)* **94**, 13850–13855.

Index